Global Climate Change

Exploring
Environmental
Challenges

**A MULTIDISCIPLINARY
APPROACH**

Global
Climate
Change

EDITED BY
Sharon L. Spray & Karen L. McGlothlin

ROWMAN & LITTLEFIELD PUBLISHERS, INC.
Lanham • Boulder • New York • Oxford

ROWMAN & LITTLEFIELD PUBLISHERS, INC.

Published in the United States of America
by Rowman & Littlefield Publishers, Inc.
4720 Boston Way, Lanham, Maryland 20706
www.rowmanlittlefield.com

12 Hid's Copse Road
Cumnor Hill, Oxford OX2 9JJ, England

Copyright © 2002 by Rowman & Littlefield Publishers, Inc.

All rights reserved. No part of this publication may be reproduced, stored in a retrieval system, or transmitted in any form or by any means, electronic, mechanical, photocopying, recording, or otherwise, without the prior permission of the publisher.

British Library Cataloguing in Publication Information Available

Library of Congress Cataloging-in-Publication Data Available

ISBN: 0-7425-2220-2 (hardcover)
ISBN: 0-7425-2221-0 (paperback)

Printed in the United States of America

♾™ The paper used in this publication meets the minimum requirements of American National Standard for Information Sciences—Permanence of Paper for Printed Library Materials, ANSI/NISO Z39.48-1992.

Contents

Contributors	vii
Preface	xi
Introduction	xv
Sharon L. Spray and Karen L. McGlothlin	

1 FROM ICE CORES TO TREE RINGS   3
Understanding Climate Change from a
Geological Perspective
David M. Dobson

2 BEYOND THE HORIZON   31
Understanding the Atmospheric Physics of
Climate Change
Richard B. Kay and Jonathan A. R. Rall

3 ECOSYSTEMS IN DANGER   59
An Ecological Perspective on Climate Change
Susan Herrod-Julius and John McCarty

4 CULTURAL PARADIGMS AND CHALLENGES   81
An Anthropological Perspective on Climate Change
Celeste Ray

5 MONEY, MARKETS, AND PRIORITIES   101
An Economic Perspective on Climate Change
Eban Goodstein

6 NEGOTIATING OUR CLIMATE 121
Understanding the Politics of Climate Change
Marvin S. Soroos

7 THE LONG ROAD AHEAD 145
Concluding Thoughts on Climate Change
Sharon L. Spray and Karen L. McGlothlin

Glossary 157
Index 167

Contributors

Dr. David M. Dobson studied as an undergraduate at Harvard University with a focus in planetary geology and did graduate work at the University of Michigan in marine geology, oceanography, and paleoclimatology. He teaches environmental geology, marine geology, and sedimentation and stratigraphy at Guilford College in Greensboro, North Carolina, where he is chair of the Department of Geology and Earth Sciences and codirector of the Environmental Studies program.

Dr. Eban Goodstein is an associate professor of economics at Lewis and Clark College in Portland, Oregon, and a research associate at the Economic Policy Institute in Washington, D.C. He received his B.A. from Williams College and his Ph.D. from the University of Michigan. Professor Goodstein is the author of a college textbook, *Economics and the Environment* (John Wiley and Sons, 1998), now in its second edition, and *The Trade-off Myth: Fact and Fiction about Jobs and the Environment* (Island Press, 1999).

Ms. Susan Herrod-Julius received her M.S. in public management and policy from the Heinz School at Carnegie Mellon University in 1989 and her B.F.A. from Carnegie Mellon in 1981. She is currently a senior environmental assessment specialist in the Global Change Research Program (GCRP) of the U.S. Environmental Protection Agency, where she is responsible for developing the ecosystem research and assessment program for the GCRP. She has also managed projects examining climate change effects on passerines, shorebirds, ducks in the prairie wetlands, and freshwater fish. She managed the development of an integrated assessment model for analyzing potential policy responses to climate change and a decision-support model for resource managers that applies the multicriteria methodology to adaptation policies.

Dr. Richard B. Kay is an emeritus professor of physics at American University in Washington, D.C. He has designed and taught courses on the environment and atmosphere in the American University Interdisciplinary Environmental Studies Program. His current research involves the design and development of lasers and laser systems for space application for NASA/Goddard Space Flight Center in Greenbelt, Maryland.

Dr. John McCarty earned his Ph.D. in ecology and evolutionary biology from Cornell University where he studied the effects of environmental variation on migratory birds. He has held the Roger Revelle Fellowship in Global Stewardship, sponsored by the American Association for the Advancement of Science (AAAS) and worked with the USDA Forest Service in Washington, D.C. on issues related to global change and the management and conservation of natural resources. He is currently an associate professor of biology and director of the Environmental Studies program at the University of Nebraska at Omaha.

Dr. Karen L. McGlothlin received her M.S. in biological sciences from East Tennessee State University and her Ph.D. in zoology from Clemson University. She is currently an assistant professor of biology at the University of the South in Sewanee, Tennessee, and teaches courses in invertebrate zoology, island ecology, entomology, and developmental biology. She also is an active participant in the interdisciplinary Environmental Studies concentration and in the Island Ecology Program.

Dr. Jonathan A. R. Rall studied physics and electrical engineering as an undergraduate, receiving B.S. degrees from both American University and Washington University in St. Louis, Missouri. He continued his graduate training at American University while working at NASA Goddard Space Flight Center in Greenbelt, Maryland, where he currently holds the position of optical physicist. His research includes developing lidar (light detection and ranging) instruments capable of measuring atmospheric water vapor, carbon dioxide, and polar stratospheric clouds. He has spent

Contributors

six summer seasons executing his research from the U.S. Amundsen-Scott South Pole Station in Antarctica.

Dr. Celeste Ray received her Ph.D. in anthropology at the University of North Carolina at Chapel Hill. She is an assistant professor of anthropology at the University of the South, where she teaches courses in cultural anthropology and environmental anthropology. She also directs an annual summer field school in historical ecology in Italy's South Tyrol through the University of New Orleans.

Dr. Marvin S. Soroos is a professor of political science and public administration at the North Carolina State University at Raleigh where he has specialized in international environmental law and policy. He is the author of *Beyond Sovereignty: The Challenge of Global Policy* (The University of South Carolina Press, 1986) and *The Endangered Atmosphere: Preserving a Global Commons* (The University of South Carolina Press, 1997).

Dr. Sharon L. Spray, an assistant professor of political science and environmental studies at Elon University in Elon, North Carolina, earned her Ph.D. from The Claremont Graduate School. In addition to teaching courses in American politics, international environmental policy, and domestic environmental politics and law, she serves as director of the Elon University Center for Public Opinion and Polling.

Preface

Exploring Environmental Challenges: A Multidisciplinary Approach is a series of short readers designed for introductory-level, interdisciplinary environmental sciences or environmental studies courses. Each reader, focused on a single, complex topic of environmental concern, outlines the concepts, methods, and current research approaches used in the study of that particular environmental challenge. Perspectives from six distinct fields of study in the natural sciences, social sciences, and humanities are represented in each volume. This approach enables students and faculty alike to become familiar with a topic from perspectives outside their own training and to develop a broader appreciation of the breadth of efforts involved in investigating select, complex environmental issues.

This series was developed to facilitate interdisciplinary teaching in environmental studies programs by acknowledging that different disciplines bring distinctly different perspectives to the table and that scholars trained in those fields are best suited to explain these perspectives. The texts in this series are designed to assist faculty trained in a traditional social science, natural science, or humanities field to venture into areas of research outside their own training.

In the last decade, a rapidly increasing number of institutions of higher education across the country have developed a wide variety of interdisciplinary programs in both environmental science and environmental studies. Although many of these programs are centered primarily within the science curriculum, more and more institutions are strengthening their environmental sciences and environmental studies majors, minors, and concentrations by adding courses from both the social sciences and the humanities. The importance of integrating information from a variety of disciplines, including the sciences, social sciences, and the humanities, has been recognized and considered in the design and revision of environmental curricula. Liberal arts institutions, in particular, are moving toward the development of interdisciplinary or multidisciplinary approaches as a basis for their environmental programs. These approaches are as varied as the institutions themselves. While many programs offer team-taught courses to provide true interdisciplinary approaches, others are built around a series of courses from across curricula that address environmental topics. The foundation for, and value of, such programs is the recognition that complex environmental challenges will necessarily require strengthening the interface among the social sciences, humanities, and natural sciences if we hope to find productive ways of addressing these challenges.

The concept for this series grew from discussions that emerged during the planning and development stages of an environmental studies program at the University of the South during the late 1990s. One of the points agreed on during our discussions was that all students enrolled in the environmental studies programs would be required to take an introductory course that would be interdisciplinary in nature and team taught by professors from the natural sciences and the social sciences. In our case, that meant a teaching collaboration between a zoologist and a political scientist.

During the course of our conversations and explorations into the available literature, we found ourselves feeling a bit overwhelmed at the thought of teaching a truly interdisciplinary course. We felt that it would be difficult, at best, to hold classroom discussions on different concepts and approaches to the study of various environmental topics from a variety of academic disciplines with our training centered in our particular fields. After much discussion, we decided that a series of edited readers, with each volume focusing on a single, complex environmental topic, with chapters written by experts in various fields, would be of great use to students and faculty involved in interdisciplinary environmental studies programs. Thus, we began the development of this multidisciplinary series of readers on environmental challenges.

Preface   **xiii**

During the conceptualization stage of this project and later, during the proposal review stage, the issue of "multidisciplinary" versus "interdisciplinary" teaching surfaced repeatedly. These two terms are frequently heard in discussions pertaining to environmental studies programs and often are used interchangeably. For this reason, we feel that it is important that these two terms be defined, in the hopes of providing clarification for some and reassurance to others that a volume written from a multidisciplinary perspective can be used in an interdisciplinary course.

When we speak of multidisciplinary perspectives we are referring to distinct disciplinary approaches to the study of a particular topic. Such perspectives do not preclude the integration of knowledge or material from other fields, but the interpretation of the information reflects a particular disciplinary perspective. We view this as a matter of disciplinary depth. As scholars we necessarily cross the boundaries of knowledge and scholarship from other fields, but most of us have more depth in the field in which we received our academic training. Consequently we interpret information through particular theoretical perspectives tied to our disciplinary training.

We view interdisciplinary teaching as the attempt at balanced integration of material from multiple disciplines. This, however, is a difficult goal when studying environmental issues. Most texts written about specific environmental issues reflect heavy bias toward the natural sciences with some discussion of policy and economics, or, alternatively, the focus may be in the opposite direction with an emphasis on policy and economics and limited discussion of science. More problematic is that many of the available texts fail to incorporate in any meaningful way the work of humanists, anthropologists, or sociologists—areas we believe are essential for understanding complex environmental challenges.

The texts are purposefully balanced with half of the chapter contributions from the natural sciences and the other half from the humanities or social sciences. Each chapter identifies important concepts and theoretical perspectives from each field and includes a supplemental reading list to facilitate additional study. We envision these texts to be the foundation for introductory environmental studies courses that examine environmental topics from multiple perspectives or other courses that seek an interdisciplinary focus for the study of environmental problems. Because we anticipate that students from a variety of majors, both science and nonscience, will use these texts, the chapters are designed to be understandable to those with little familiarity of the topic or the field about which it was written.

The series is not neutral in its basic premise. The various topics in the series were chosen because we believe that the topics addressed are environmental challenges that we want students to better understand and hopefully work toward future solutions. Individual authors, however, were asked to provide objective presentations of information so that students and faculty members could form their own opinions on how these challenges should be addressed. We care deeply about the environment and we hope that this series serves to stimulate students to take the earth's stewardship seriously and promote a better understanding of the complexity of some of the environmental challenges facing us in this new century.

Sharon L. Spray
Karen L. McGlothlin

Introduction

Sharon L. Spray
and
Karen L. McGlothlin

In the early 1990s, oceanographer Dr. James J. McCarthy, director of Harvard University's Museum of Comparative Zoology, accompanied a group of tourists aboard a Russian icebreaker to the North Pole. On this journey, the ship plowed its way through six to nine feet of ice to reach the top of the earth. When Dr. McCarthy returned to the same location with another group in August 2000, he and the group were surprised to find a mile wide stretch of ice-free water. Although approximately 10% of the Arctic Ocean may be free of ice in a typical summer, this report of open water at the pole has been interpreted by some scientists as another piece of evidence to add to the growing list of events related to global warming and climate change.

In the United States, Glacier National Park in Montana, home to more than fifty glaciers and vacation destination of more than 1,686,000 people in 1999, may not be attracting the tourist crowds in the near future. Mark F. Meier, in the Walter B. Langbein Memorial Lecture at the spring 1998 meeting of the American Geophysical Union, estimated that no glaciers will remain in Glacier National Park by 2070 if glacial retreat continues at the current rate.

These are only anecdotal pieces of evidence and would not alone be considered strong scientific evidence of climatic change. But as the chapters in this volume indicate, there is a strong body of scientific research that leads us to conclude that our planet is undergoing an unprecedented period of warming along with global climatic changes that may forever alter our planet.

This volume is meant to strengthen the reader's understanding of the scientific basis for the mainstream scientific community's understanding of global warming and current climatic changes. Global climate change, as an environmental challenge, is an extraordinarily complex issue that involves integrated knowledge drawn from various fields of study. It is not solely a scientific issue that can be approached from a single field, such as geology, nor is it solely a matter of public policy or economics. To meet the challenge of a changing climate, it is imperative that we understand the issue on as many levels as possible. Our success will be influenced by our understanding of current science, historical responses to past environmental problems, and numerous economic and policy choices.

This collection of essays is designed to introduce readers to the topic of climate change from multiple disciplinary perspectives. It is not designed to promote one policy prescription over another, or to call readers to action. Yet this book is not value neutral. It begins with the premise that global warming is occurring and that climatic change poses a threat to our natural environment and the health of future generations. We approach the topic with the belief that the more knowledgeable people are about the topic, the better equipped they will be to address climate change.

Introduction

The chapters in this volume are designed as introductions to the topic, so many of the concepts are only broadly covered. The suggested reading lists that accompany each chapter will facilitate more in-depth study from the perspective of each of the represented disciplines. The order of the chapters is not meant to suggest that one disciplinary perspective is more valuable to the discussion than any other; rather, the order reflects how environmental problems come to be recognized as problems—by first identifying and documenting a disturbance in the natural environment and then following up with a discussion of responses. Each chapter may be used as a stand-alone unit or may be read in conjunction with others. For this reason, there is some basic informational overlap from chapter to chapter that allows nonsequential reading. However, we encourage readers to start with the chapters that address the scientific data before they look at how global climate change is being studied by those outside the natural sciences.

Included in this book are six chapters written from six very different disciplinary perspectives. Chapter 1, "From Ice Cores to Tree Rings: Understanding Climate Change from a Geological Perspective," introduces the reader to the historical scientific record of climate change and how those changes are investigated and documented. It is important to establish whether the pattern of warming that we are observing at this time is part of a larger, natural cycle of climate change or whether there are differences in the current rates and patterns of warming that might indicate otherwise. The study of climate change from a geological perspective may enable us to determine this. This chapter is followed by "Beyond the Horizon: Understanding the Atmospherics Physics of Climate Change," in which the authors explain the atmospheric interactions that are occurring at this time and how those interactions contribute to our changing climate. Once the historical stage has been set and the reader has been introduced to the mechanisms involved in the current climate change, chapter 3, "Ecosystems in Danger: An Ecological Perspective on Climate Change," describes the potential effects of climate change on the distribution and abundance of a variety of organisms. The examples in this chapter introduce the ecological concepts that are relevant to climate change and describe some of the ways in which organisms may adapt to that change.

Whether someone lives in a highly industrialized society or in a remote tribal setting, a society's cultural beliefs, the values a society places on the natural environment and the resources at its disposal will shape future responses to climate change. The social science chapters begin with chapter 4, "Cultural Paradigms and Challenges: An Anthropological Perspective on Climate Change," which discusses human adaptation to past climatic changes. Chapter 5, "Money, Markets, and Priorities: An Economic Perspective on Climate Change," introduces the economic concepts and policy options currently on the table in international negotiations and critical for understanding current policy discussions and mitigation strategies to address global warming. The final disciplinary perspective, chapter 6, "Negotiating Our Climate: Understanding the Politics of Climate Change," discusses the current state of international cooperation to address global warming and the political dynamics of the United States' response to this highly complex, transboundary environmental problem.

We realize that the study of climate change encompasses additional disciplines; however, not all could be included. This is a reflection of space limitations, not our opinion of which fields are important. We hope that this volume will introduce students to the topic from those disciplines included and serve as a springboard to explore the approaches of others.

Global Climate Change

CHAPTER

1

From Ice Cores to Tree Rings

UNDERSTANDING
CLIMATE CHANGE
FROM A
GEOLOGICAL
PERSPECTIVE

David M. Dobson

Fossil fuels, carbon dioxide, the greenhouse effect, global warming—all of these are concepts tossed about in the media, negotiated by politicians, joked about by late-night comedians, and pondered by people worldwide. Science can observe, analyze, and make predictions about almost all subjects, and global climate change is no different. Climate change is one of the key areas in which the scientific observations and analyses could have a profound impact on how people everywhere live their lives.

Geologists spend most of their time trying to understand earth history and thus are particularly interested in global climate change. Many of the most fascinating periods of geologic study revolve around climate and climate change. Some of these events happened millions or billions of years ago—the origin of life, the formation of the oceans, the advent of atmospheric oxygen and thus animal life, and the extinctions at the end of the Permian period (245 million years ago, when 95% of marine species went extinct) and at the end of the Cretaceous period (66 million years ago, which paved the way for mammal diversification). Some happened only a few million years ago, such as the recent set of massive ice ages and interglacial warm periods, roughly coincident with the evolution of our own species. Some even occurred during historic time—climate and climate change have been implicated in the multimillennial success of the Egyptian culture, in the rise and fall of Chinese dynasties, in the spread of Viking exploration, in the Renaissance, and in the development and disbanding of the Anasazi cliff-dwelling culture. The study of global climate change offers an exceptional opportunity to apply what we know about earth history to a pressing modern problem.

Some fields within earth science are more closely focused on climate change than others. Climatologists study how sunlight, winds, water currents, pressure systems, and landforms interact to produce long-term weather conditions. Most climatologists use computer models based on modern instrumental measurements to analyze and predict climate change. Paleoclimatologists often also make use of computer models, but they use other techniques to go further back in time, studying geologic records such as ocean or lake sediments, ice cores, oxygen isotopes, and coral growth bands to learn how climate has varied in the past. It takes sedimentologists, glaciologists, geochemists, and paleontologists to produce these records. Oceanographers study short- and long-term behavior of the oceans, and the oceans are always a dominant force in climate.

The key to understanding geological research (and thus, **paleoclimatology**) is to remember a couple of characteristic aspects of geology. Geology is often more interpretive and historical than are other sciences. Geologists are routinely faced with imperfect and cryptic evidence of events long past and must struggle to create hypotheses about what might have happened or what forces might have been at work. Geology is also very integrative—most geologists are broadly trained and need to combine elements of chemistry, physics, biology, mathematics, and history in their research. The study of climate change (past and present) includes all of these aspects.

Geologic processes run in nature rather than in the laboratory, and climate change is no exception. We are in the midst of a giant global experiment, carried out by everyone everywhere, from a paper mill manager in Oregon to a family subsisting through slash-and-burn in the Amazon basin, to a cowboy in Texas moving cattle to greener pastures, to a rice farmer in Laos tending her paddies. The test tube for this experiment is the earth's atmosphere, and we are all contributing ingredients. The key is to figure out what the result of our experiment might be in time to stop it if we need to—if we can indeed stop it, and if it is not too late to do so.

Scientific Concepts

To start, let's get our terms straight, beginning with the result of the experiment and moving backward to the ingredients.

GLOBAL WARMING

Global warming is what we are worried about, and it is exactly as it sounds—a trend toward warmer conditions around the world. Part of the warming is natural; we have experienced a 20,000-year-long warming as the last ice age ended and the ice melted away. However, we have already reached temperatures that are in line with other interglacial (minimum ice) periods, so continued warming is likely not natural. We are contributing to a predicted worldwide increase in temperatures ranging between 1° and 6°C (2° and 10°F) over the next 100 years. The warming will be more dramatic in some areas, less in others, and some places may even cool off. Likewise, the impact of this warming will be very different depending on where you are—coastal areas must worry about rising sea level, while Siberia and northern Canada may become more habitable and appealing for humans than these areas are now.

6 David M. Dobson

The fact remains, however, that it will likely get warmer, on average, everywhere. Scientists are in general agreement that this process has already begun and that the warmer conditions we have been experiencing at the end of the twentieth century are at least in part the result of a human-induced global warming trend (Intergovernmental Panel on Climate Change [IPCC] 1995). Some scientists maintain that the changes we are seeing fall within the range of random variation—some years are cold, others warm, and we have just had an unremarkable string of warm years recently—but that is becoming an increasingly rare interpretation in the face of continued and increasing warm conditions.

GREENHOUSE EFFECT

Why will this warming occur? It will occur because of the **greenhouse effect**. The greenhouse effect is one of the memorable analogies that geologists are very fond of using, but in this case, it is a very good match. Sunlight passes through the glass walls and ceiling of a greenhouse and strikes plants, pots, and dirt. Some of the light is reflected, and it goes back out through the glass. Some of the light is absorbed, and when it is absorbed, it heats up whatever it is striking. Here is where a key concept comes into play—*hot objects give off energy.* The sun is a really hot object, so it gives off mostly visible light. The objects in the greenhouse are not nearly hot enough to give off visible light (as molten iron can, for example), but they do give off infrared light (heat). The glass in the greenhouse is transparent to visible light, but it is not transparent to infrared light. This means the infrared light, or heat, is trapped in the greenhouse, and the greenhouse heats up.

This effect happens everywhere you have sunlight and glass—in a car left in a parking lot on a hot day, or in a sunny room with a big picture window. The reason this analogy works for global warming is that the earth's atmosphere acts a lot like the window of a greenhouse or car. Sunlight (visible light) passes through the earth's atmosphere, striking the earth's surface. Some of the sunlight is reflected back into space; some is absorbed. The light that is absorbed heats up the earth's surface. As a warm object, the earth gives off infrared radiation. Some of this infrared radiation escapes into space, but some is absorbed by the atmosphere.

GREENHOUSE GASES

The atmosphere consists mostly of nitrogen (78%) and oxygen (21%). Neither gas absorbs infrared radiation, nor do most components of the atmosphere. However, some components, called **greenhouse gases** (**GHGs**), do absorb infrared light. The most abundant greenhouse gas is **carbon dioxide**, commonly referred to by its chemical formula, CO_2. This gas makes up under one-half of one percent (0.5%) of the earth's atmosphere—not too much, but important for the delicate global heat balance. Another important greenhouse gas is natural gas, or **methane** (CH_4). Methane is much less common in the atmosphere (around 1,700 ppb [parts per billion], or $1/200\text{th}$ as common as CO_2), but it is much more effective—about twenty-one times as effective—at absorbing infrared radiation than carbon dioxide. These two are the major players in human-induced climate change; there are other greenhouse gases (e.g., **chlorofluorocarbons**, or **CFCs**), but they are either less effective or less abundant than CO_2 and methane or controlled primarily by natural forces (e.g., water vapor).

FOSSIL FUELS

Why do we care about the greenhouse effect now? A good question—we have had CO_2 in the atmosphere as long as we have had an atmosphere, so there has always been a greenhouse effect. In fact, with no greenhouse effect, the surface of the earth would be as much as 33°C (60°F) colder. The problem is one of magnitude. Over the past two or three centuries, humans have started adding more greenhouse gases to the atmosphere. Although there are many sources of these new greenhouse gases, the primary source is **fossil fuels** used for industry, heating, transportation, and electricity. A fossil fuel is a substance found in the earth that is formed over a long time from ancient life. Common fossil fuels are coal, natural gas, and oil (which can be refined into gasoline). Fossil fuels contain high concentrations of carbon, and it is the burning, or oxidation, of carbon that releases the energy we use. Written as a simplified chemical formula, it looks like this:

$$C_{(fossil\ fuel)} + O_{2(atmosphere)} \rightarrow CO_{2(atmosphere)} + heat\ energy$$

You can see that one of the products of burning fossil fuels is CO_2, a greenhouse gas. This means that burning fossil fuels adds CO_2 to the atmosphere. More CO_2 in the atmosphere means that more heat is being absorbed rather than being radiated back into space, and the atmosphere (and thus the surface of the earth) gets warmer.

Why has this become a problem just over the past few centuries, when people have been burning things to keep warm for more than 500,000 years? It is true that burning wood or leaves or animal fat produces CO_2 just like the fossil fuels. The key here is that the materials we are burning now are true fossils—they are the remains of organisms that lived long ago—sometimes 200 million years or more. If you cut down a tree and burn it, you are only really recycling carbon. The tree, as it grew, absorbed CO_2 from the atmosphere and, through **photosynthesis**, converted that CO_2 to carbon that it could use in sugars, amino acids, proteins, and other materials. When you burn it, you are returning that carbon to the atmosphere, and it has been stored in the tree for probably only a few decades at the most.

When you burn a fossil fuel, you are taking carbon out of the ground, where it has been trapped and inactive for millions of years, and releasing that carbon to the atmosphere, where it is, in a sense, "excess" carbon. Before the **Industrial Revolution**, humans (who were present in much smaller numbers than today) just burned things that had died recently, recycling CO_2 over the short term. By digging up fossil carbon and burning it, we have started adding more CO_2 to the atmosphere than the existing plants and forests are accustomed to removing, so the concentration of CO_2 in the atmosphere has gone up. Population growth is also a major contributor—as human populations have increased exponentially, so too has our use of combustible fuel and our production of CO_2.

Feedback

Sometimes, a change in a system sets in motion other changes that either reinforce or weaken the original change. This is called **feedback.** If you put a microphone in front of a speaker connected to the microphone, you can hear an electronic example of feedback—the hum of the speaker system is picked up by the microphone, which relays the sound to the speakers, which produce that sound plus their original hum, which is again picked up by the microphone and sent back to the speaker. The original hum is amplified many times as this cycle continues, and that eventually produces a loud screech. This reinforcing and amplifying is called positive feedback, while forces that tend to damp out or weaken a change are called negative feedback.

The climate system works the same way. For example, consider rocks, soil, and vegetation, and compare them with ice. Ice is white or light gray—a very reflective, bright color. Rocks, soil, and vegetation are darker grays, browns, and greens—not as reflective. The reflectivity of a material is called its **albedo** (see Kay and Rall, this volume), which is defined as the percent of incoming light that is reflected by a substance. So, ice has a high albedo, and the other materials have relatively lower albedos. Another way albedo can have a strong effect is through volcanic eruptions—dust and ash blown into the air by volcanoes can reflect sunlight before it ever reaches the earth's surface, cooling the planet. This is a major (if usually short-lived) process; eruptions like those at Mount Pinatubo in the Philippines can cause a global cooling of several degrees for a year or more.

Why care about albedo? Albedo provides a sizable positive feedback to the climate system: The colder it gets, the more ice will form; the more ice that forms, the higher the albedo of the earth's surface will be. The higher the albedo, the more incoming sunlight will be reflected back to space (remember that visible light is not trapped by CO_2 or other greenhouse gases), and the less it will heat the earth, making the earth even colder. So, an initial cooling is reinforced, leading to further cooling—a positive feedback loop (figure 1a).

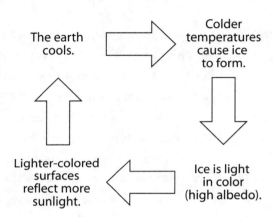

Figure 1a. When cooling leads to the formation of ice, its high albedo causes increased cooling (a positive feedback loop).

Note that the same positive feedback loop would also reinforce a warming trend. A warming trend would melt ice, decrease albedo, increase the amount of sunlight absorbed, and lead to more heating (figure 1b).

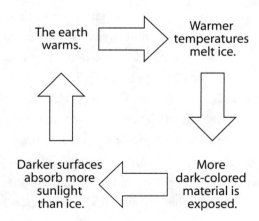

Figure 1b. When warming results in melting of ice, the decreased albedo can lead to enhanced warming (also a positive feedback loop).

There is also negative feedback and **buffering** in climate systems. For example, an increase in atmospheric CO_2 is partially compensated by the ocean—when CO$_2$ is added to the air over the ocean, the rate that CO$_2$ dissolves into the ocean increases, eliminating some of the effect of the increase. Another example is clouds—higher temperatures will tend to increase evaporation from the ocean, which puts more moisture in the atmosphere, which can lead to more clouds. Clouds have a high albedo, and they reflect sunlight, although they can serve as a greenhouse gas, trapping heat radiating from the earth (figure 2).

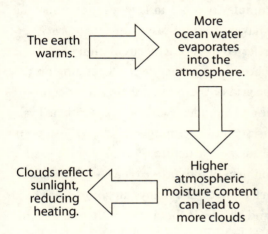

Figure 2. Warming can also cause negative feedback or buffering, which resists or counteracts an initial change. Note, however, that water vapor is a greenhouse gas, so increased atmospheric water content might also provide positive feedback.

Feedback and other climatic responses are very important in understanding the climate, because they will often affect, or even dictate, the extent of climate changes. We are able to estimate and even measure with fair accuracy how much carbon we are adding to the atmosphere, but without understanding the feedback mechanisms that respond to that change, we cannot predict the result. If positive feedback dominates, much more heating may occur than just the heat trapped by our excess CO$_2$; if negative feedback dominates, we may see very little change.

GLOBAL CLIMATE MODELS

Weather prediction is a notoriously difficult business, even for weather that is only a day or two away. In fact, it is the study of weather prediction that led in part to the development of chaos theory—there is so much complexity to weather patterns that we cannot expect to know or model exactly what will occur or when.

Prediction of climate change is in some ways more difficult, because it requires predictions much further into the future, but it is simpler, too, because the small day-to-day or year-to-year variations can be ignored in the pursuit of the overall trend. Most climate predictions are made using sophisticated global climate models running on computers. Given current computer limitations and the complexity of the task, we cannot model the entire world's climate without making compromises.

Most climate models (often referred to as **general circulation models**, or **GCMs**) divide the planet's surface into a grid system. Depending on the sophistication of the models, the grid cells can range from a few degrees to fractions of degrees of longitude and latitude. Each grid cell is assigned values for a number of parameters, such as incoming sunlight (adjusted for seasons), atmospheric composition and structure, elevation of the earth's surface, composition and environment of the surface (at a coarse level, land or water; at a sophisticated level, there can be many categories from lush jungle to arctic ice to desert), and albedo of the surface. Often, the entire vertical air column is modeled, including wind speeds, high and low pressure systems, and precipitation.

The models also contain rules for how heat, air, and water behave in a given cell, and how changes in one cell affect changes in its neighbors. The rules are based on previously observed behavior of the atmosphere. For example, warm land areas will tend to heat up air, causing it to rise; air from surrounding areas will enter at low elevations to replace the rising air. The rising air will tend to cool, causing the water vapor it contains to condense into clouds, rain, or snow. All of these behaviors can be incorporated into a GCM. The GCMs are also frequently updated whenever a new set of observations, a new understanding of atmospheric processes, or a better computer system is developed.

Once the grid has been created, the parameters have been set, and the rules devised, the model is run through a series of discrete time intervals.

From Ice Cores to Tree Rings

Depending on the scope of the model, the time intervals can be anywhere from hours to thousands of years. The model is a simulated planet, and as the time intervals pass, this planet goes through changes based on the starting conditions and rules it contains. The resulting conditions on the planet at the end of the run become a prediction of what might happen on the real earth. One of the most useful aspects of climate modeling is that you can specify hypothetical conditions that have not yet occurred—for example, many global warming predictions are based on climate models in which CO_2 is increased over time to a concentration double that of the modern atmosphere.

How good are the predictions of GCMs? That is a difficult question to answer, but a very important one. The accuracy of the predictions depends on the construction of the model and the importance assigned to the different parameters. A model's output can vary wildly even with small changes in some starting conditions or rules while remaining indifferent to changes in other parameters. One of the only ways to test these models is to have them simulate a time period that we know a great deal about. An example could be the climate of the past 100 years, which we know very well from instrumental records, or for a longer-term study, the climate change through the latter stages of the most recent ice age, which we know pretty well from natural records.

Often, modelers cannot give specific numbers for predictions of change, but they can with fair confidence predict a range of possible outcomes based on repeated model runs with varied starting conditions and rules. Modeling is a tricky business, but an important one for predicting climate.

FLUX MODELS

To understand a system in which quantities of substances are important, geoscientists construct **flux models** to study where substances travel and how long they stay. Flux models usually consist of one or more **reservoirs**, or places where the substance resides, and one or more **sources** and **sinks** for each reservoir. Sources are processes that bring the substance into a reservoir, whereas sinks are processes that remove a substance from a reservoir. For example, figure 3 shows a simple flux model of the ocean and its water supply.

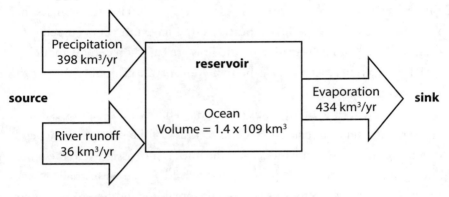

Figure 3. Simple flux model for water in the ocean.

In this model, the ocean is the reservoir, precipitation and runoff are sources, and evaporation is a sink. If the sources are equal in magnitude to the sinks, as they are here (398 + 36 = 434), then the system is in **steady state**. This means that the reservoir will neither grow nor shrink, because the same amount is being added that is being taken away. Many natural systems appear to be in a steady state unless they have recently been disturbed, usually by human activity. However, atmospheric CO_2 is not in steady state at the moment—more CO_2 is being added than is being removed, so the atmospheric reservoir of CO_2 is growing.

One more concept to know about is **residence time**. Residence time is the average amount of time any given unit of a substance will reside in a particular reservoir. It is calculated like this:

$$\text{residence time} = \frac{\text{reservoir size}}{\text{total sinks}}$$

Note that if the system is in steady state, the total sinks will be equal to the total sources. For the ocean example, the residence time is:

$$\text{residence time} = \frac{1.4 \times 10^9 \text{ km}^3}{434 \text{ km}^3/\text{yr}} = 3.2 \times 10^6 \text{ yr}$$

So, the average water molecule will spend about 3 million years in the ocean before being evaporated and removed from the ocean. The residence time of CO_2 in the atmosphere is very important to global warming, because it determines how long the excess CO_2 we are adding will stay around and heat the earth.

Constructing flux models is sometimes complicated. Geologists studying carbon cycling need to use several reservoirs where carbon resides—in the atmosphere, which is the greatest concern for global warming, but also in biological organisms, in soils, in rocks and sediments, in fossil fuels, and in seawater. They need to estimate how much carbon is stored in these reservoirs. They also need to know how much carbon is transferred from one reservoir to another over time, and they need to make sure that they are accounting for all such transfers.

Climate Proxies

One of the challenges in reconstructing past temperatures is that prior to about 1900 there were very few good long-term instrumental records of temperature being created, and even the records that do exist seldom go back to before 1700 or so. Geoscientists are used to this kind of challenge, however, and their solution is to look for **climate proxies**—natural data sets that reveal aspects of the climate in which they were formed. Tree rings are a good example—their timing is annual and obvious, and their thickness and other measurements can reflect both temperature and rainfall patterns in the area where a tree grew. Annual growth bands in corals are likewise a good climate record. The thickness of annual ice deposits or records of ice melt events can also be very useful.

One of the main proxies geologists use to examine past temperatures and climate is the analysis of **stable isotopes**. Isotopes of a particular element all have the same number of protons (thus the same atomic number) but have differing numbers of neutrons. For example, most oxygen atoms have eight protons and eight neutrons, for a total atomic weight of sixteen (written as ^{16}O), but a small fraction have ten neutrons for a weight of eighteen (written ^{18}O), and some even have nine neutrons (^{17}O). Most chemical elements have several different stable isotopes, and some elements have unstable radioactive isotopes (e.g., ^{12}C and ^{13}C are stable isotopes of carbon, but ^{14}C is radioactive).

Any isotope of a particular element can form a compound with other elements in much the same way. For example, water, H_2O, can be made with ^{16}O, ^{17}O, or ^{18}O, and it can be made with regular hydrogen (^{1}H) or with deuterium (^{2}H) or tritium (^{3}H—a radioactive isotope). The chemical properties of this water molecule will be almost identical regardless of which isotopes are included. However, the reason stable isotope ratios are important as climate proxies (and for other research purposes) is that there can be a subtle difference in the behavior of different isotopes. For example, water containing ^{18}O is slightly less likely to evaporate than water containing ^{16}O, so over time, atmospheric water vapor will contain relatively less ^{18}O compared with ^{16}O than the ocean. Similarly, because of slight differences in chemical bonding, the ratio of ^{18}O to ^{16}O in shells of marine organisms is different from that of oxygen in seawater. This difference in ratios is called **isotopic fractionation**—a sorting of isotopes by weight in different compounds.

How does this apply to climate? There are two main applications. The first is as a proxy for ice volume. When ice sheets form on the continents, the ocean becomes more enriched in ^{18}O. This is because the atmosphere, where rain and snow come from, is enriched in ^{16}O relative to the ocean. When ice forms, it locks up ^{16}O that would otherwise return to the ocean in rivers. This makes the ocean even more enriched in ^{18}O than it already is. Thus, if you can find an old shell and measure its ratio of ^{16}O to ^{18}O, you can often get information on how much ice was around when the organism that made the shell was living.

The second application involves temperature. In many cases, the extent of isotopic fractionation, particularly oxygen isotope fractionation, is dependent on temperature. Usually, the warmer it is, the less fractionation occurs. Thus, if shells form in cold water, they will tend to have more ^{18}O than if they formed in warm water. This means you can learn about past temperatures by measuring relative stable isotopic concentrations (usually oxygen, but also hydrogen and carbon) of a lot of different materials. Shells and other materials made of calcium carbonate (also known as $CaCO_3$, calcite, or limestone) are the most commonly used, but bones, teeth, trees, soils, sediments, minerals, ice, seawater, and many other materials are also studied.

Current Research in the Field

There has been a great deal of controversy over the greenhouse effect and global warming issue. Some scientists and politicians have claimed it is a very serious problem requiring prompt corrective action. Others have maintained that it is highly overrated and will not produce serious consequences. Still others see it as a political or financial opportunity to promote cleaner alternate energy sources and to reduce dependence on fossil fuels. Over the past decade, though, much of the controversy has been resolved by observations and measurements that show continued warming. The general scientific consensus now is that human-induced global warming is real and will produce profound changes in our world (IPCC 1995).

The problem can be investigated scientifically, but the solutions are social and political. Because any action will be based on predictions, there are of course uncertainties and estimates to be taken into account.

ATMOSPHERIC CO$_2$—PAST AND PRESENT

Atmospheric CO$_2$ levels have been monitored in many locations over the past forty years and sometimes for even longer. These instrumental records show a steady increase in atmospheric CO$_2$ concentration over that time—an increase of nearly 15%. One of the most famous CO$_2$ data sets is the Mauna Loa record from Hawaii (figure 4).

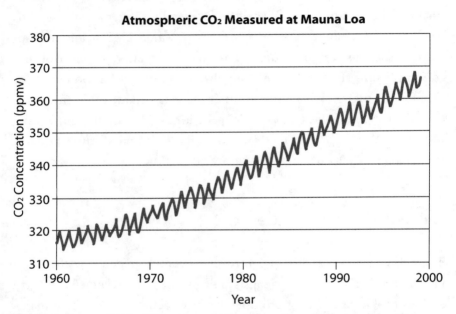

Figure 4. Atmospheric CO$_2$ concentrations over the past forty years at Mauna Loa. The small fluctuations represent normal annual variation in CO$_2$ concentrations. Data from Keeling and Whorf (1999).

We can explore further back in time by using **ice cores**, which contain frozen bubbles from past years. An example of one of these records, taken from the **Vostok** ice cores in Antarctica, is presented in figure 5. This record shows the normal pattern of variation from interglacial time, through the slow development of an ice age, through the relatively rapid melting of the most recent ice age over the past 20,000 years, to the modern warm interglacial period we now enjoy. Notice that the maximum interglacial (warm) CO$_2$ levels (270–300 ppmv [parts per million by

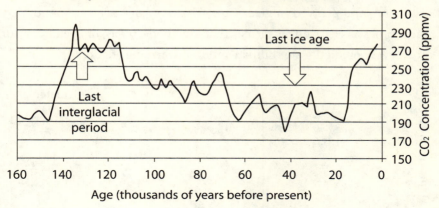

Figure 5. Atmospheric CO_2 concentrations from bubbles trapped in the Vostok ice cores, drilled in Antarctica. The period covered includes the last interglacial period, the last ice age, to the current interglacial period; CO_2 concentrations correlate very well with temperature change and ice cover. Modern atmospheric concentrations have topped 370 ppmv CO_2. Data from Barnola et al. (1987); the record is extended in Petit et al. (1999).

volume]) are still much lower than any of the modern values in figure 4 (310–370 ppmv).

These records show how much higher modern CO_2 levels are than those over the past 150,000 years and than those earlier in the twentieth century. The question remains, however, about whether these elevated CO_2 levels are causing warmer than average temperatures. In the case of the glacial-interglacial record, high CO_2 is definitely correlated with warm periods, but it is not clear from these records that the CO_2 is definitely causing the temperature increases—it could be that the CO_2 is merely a by-product or amplifier of warmer times rather than the driving force.

GLOBAL TEMPERATURE—PAST AND PRESENT

Throughout the earth's history, there have been warm periods and cold periods. There have been several major periods of widespread global glaciation, and we are in one of these periods now—in fact, our species originated and evolved in glacial times. There have been tremendously warm times as well, much warmer than the modern day, or even than our most severe predictions of human-induced warming (figure 6).

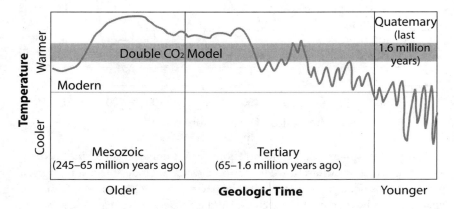

Figure 6. Relative temperature during the last 245 million years. The line marked Modern corresponds to temperatures in the year 1900, before the twentieth century warming trend. The band marked Double CO_2 Model corresponds to predictions of potential global temperature from a doubling of preindustrial atmospheric CO_2 levels. Note that the horizontal scale changes from period to period and that there is more detail in the more recent time intervals—that is not merely because there were severe climate changes (although there were), but also because we have better records of these recent times. Based on a figure prepared by Eddy and Bradley (1996).

On a more recent scale, ice cores can give us information on climate over the past few hundred thousand years (Petit et al. 1999). The Vostok ice cores, in particular, provide a record of temperature and CO_2 at the same time (figure 7). The CO_2 is measured from bubbles (see also figure 5), and the temperature comes from a stable isotope fractionation calculation from the ratio of deuterium (^2H) to regular hydrogen (^1H) in the ice. The records are derived independently of each other, yet they show a very strong correlation between high CO_2 concentrations and warm times. It is evidence like this that makes scientists confident that CO_2 is related to and may contribute to significant climate change.

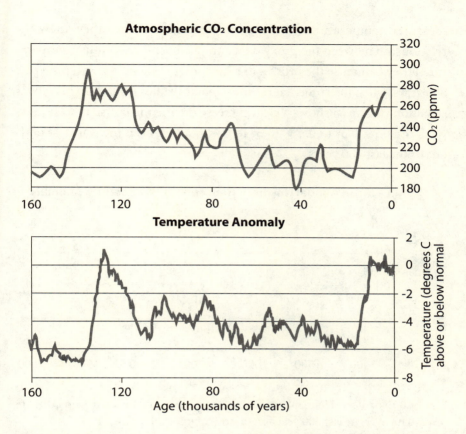

Figure 7. Comparison of CO_2 and temperature over the past 160,000 years from the Vostok ice cores. Correlation between the two records is generally very good, particularly for glacial and interglacial maxima. CO_2 data from Barnola et al. (1987); temperature data from Chappellaz and Jouzel (1992). See also Jouzel et al. (1993) and Petit et al. (1999).

As global warming has become a more major concern, geoscientists have worked hard to come up with evidence for or against warming. One group of researchers (Mann et al. 1998) has come up with compelling evidence for recent warming. These scientists assembled a great many climate proxy data sets, including tree rings, coral growth bands, oxygen isotope records, and others, and also instrumental records. Through statistical analysis, they consolidated and correlated all of these records with measured temperatures in the twentieth century. Once they knew the relationship between the proxy data and real twentieth-century temperatures, they could use the proxies to reconstruct temperatures over the past 600 years, when instrumental records are not readily available. This reconstruction (figure 8) shows a clear increase during the last sixty to seventy years of about 0.3° to 0.4°C (0.5° to 0.7°F) above even the warmest periods of the last 600 years.

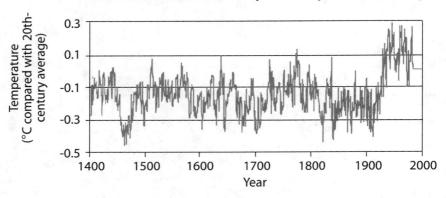

Figure 8. Northern Hemisphere temperatures as reconstructed from climate proxy data over the last 600 years. Data from Mann et al. (1998).

It is clear from this data set that the warmer temperatures we have seen at the end of the twentieth century are abnormal—not just unusual, but completely out of character for the past 600 years. That they correspond in timing and in magnitude with increased concentrations of CO_2 (a known greenhouse gas) in the atmosphere is compelling evidence that global warming is real and human induced.

CARBON FLUX MODELS

While climate models (based on geologic history and climate proxies) are the most effective tools we have to predict future climate, carbon flux models are among the most useful tools in deciding what to do about it.

The flux model shown in figure 9 shows the excess carbon being produced, about 7.4 gigatons of carbon per year (GtC/yr) from fossil fuel burning and cement production, and about 1 to 2 GtC/yr from changing land use, which includes activities such as deforestation and agriculture. Cement production releases CO_2 as limestone ($CaCO_3$) is converted to calcium oxide (CaO) at high temperatures. This process, plus the fossil fuels used in the production process, can make up as much as 2% of annual CO_2 emissions (Wilson 1993). Deforestation produces atmospheric carbon in two ways—the wood from forests is normally burned, and because forest ecosystems are more complex and dense than the

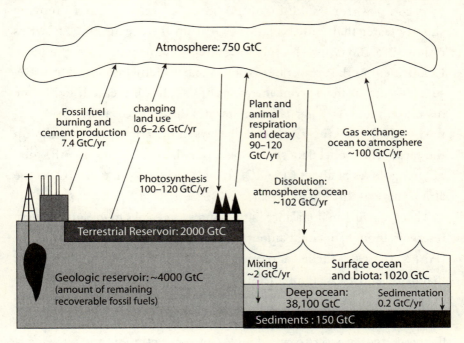

Figure 9. Global reservoirs and fluxes of carbon. All reservoirs for carbon (e.g., atmosphere, ocean) are shown in billions of tons (gigatons) of carbon (GtC). All fluxes are in gigatons of carbon per year. Note the massive storage of carbon in the deep oceans, and the imbalance in the atmospheric fluxes caused by fossil fuels. Carbonate rocks (created over time by sedimentation in the ocean) are another large reservoir that stores carbon for very long periods. Global carbon cycle data from Post et al. (1990), Mitchell and Geerts (1997), and U.S. Environmental Protection Agency (EPA) (2000).

agricultural or residential areas that replace them, there is usually less photosynthetic activity after removal of forests. Some municipal and agricultural practices are likewise more carbon intensive than others; landfills and rice paddies produce significant amounts of methane as do herds of cattle and other ruminant animals, which produce methane as a part of their digestion.

Another interesting aspect of the flux model is that it reveals some of the natural system and feedback loops in place. Photosynthesis, whereby plants remove CO_2 from the atmosphere, counters almost perfectly the respiratory activities of all living things and the decay of organic matter by decomposers. Dissolution of CO_2 into the ocean is actually moving slightly faster than outgassing of CO_2 from the ocean (102 GtC/yr as opposed to 100 GtC/yr)—this is a purely chemical process; the more CO_2 present in the atmosphere above the ocean, the more will dissolve, so as the atmospheric concentration of CO_2 has increased, so has the dissolution rate. This partially ameliorates the effects of fossil fuel burning—not all fossil fuel CO_2 actually stays in the atmosphere to heat up the earth. The model data show a reservoir size of 750 GtC and the total outgoing fluxes of about 210 GtC, giving a residence time for CO_2 in the atmosphere of three to four years.

There remains an imbalance in the sources and sinks to the atmospheric reservoir, though, one that amounts to an increase of about 2 to 5 GtC/yr (Post et al. 1990; Mitchell and Geerts 1997). Many of the climate models that geoscientists use to predict future climate assume CO_2 levels double or triple those of the year 1900. This seems extreme, but it is not. Most geologists agree that during the later parts of the Paleocene (65–57 million years ago), we were near a double CO_2 concentration, and by increasing atmospheric CO_2 by 0.5% to 1% per year, we could again reach double CO_2 within the next 200 years. As mentioned earlier, we have already seen CO_2 concentrations go up by almost 15% in the past forty years.

One final thing to note—the total geologic reservoir of fossil fuels is estimated here at about 4,000 GtC. This is a difficult figure to estimate, because it must include not only the resources we have now discovered but also those that we are likely still to discover, and it must take the

From Ice Cores to Tree Rings **25**

accessibility and cost of recovery of those materials into account. There are estimates that place our potential reserves much higher and lower than 4,000 GtC. Also, the great bulk of our remaining fossil fuels is coal reserves, which must be recovered, processed, and used differently than oil and natural gas.

Regardless, at projected rates of use, we are likely to run out of all fossil fuels before the next millennium, and oil and natural gas within the next few hundred years if not sooner. In some ways, this is very discomforting—we as a culture have built much of our industry and our wealth on fossil fuels, and no industrialized society has ever existed that did not rely on them. In another, more geologic sense, however, it is comforting—the imbalance we are now creating in the carbon cycle could go away very soon in geologic terms; the planet will reach a new global equilibrium as the reservoirs and fluxes shift around, and global warming is unlikely to threaten the long-term health of our species.

CONSEQUENCES

Geologic history gives us perspective, climate proxies and instrumental measurements give us data, and flux models give us guidance in analyzing potential climate change. All of these factors are used in the design and execution of climate models. What, then, do the climate models tell us? There are many predictions, but those that include a range of possibilities are perhaps most useful (EPA 2000). Within a reasonable margin of uncertainty, we are likely to see the following:

- CO_2 emissions, now about 7.4 GtC/yr, will likely rise to 9 to 15 GtC/yr by 2025 due to increased global population and industrialization. After that, economic models indicate they could remain constant, decline slightly, or increase to as high as 38 GtC by 2100.
- Atmospheric CO_2 will likely reach the "double CO_2" mark of about 700 ppm [parts per million] by 2100.
- Double CO_2 could increase the earth's average surface temperature by 1.5° to 4.5°C (2.7° to 8.1°F), but this will likely be lessened in the

short term through heat absorption by the oceans. By 2100, 1° to 3°C (1.8° to 5.4°F) is more likely. This may not seem like much, but remember that the global average temperature during the last ice age was only 5 degrees Centigrade (9 degrees Fahrenheit) cooler than today.

What changes will a warmer world bring? There would obviously be shifts in the climate of areas or regions, but it is difficult to predict these on a case-by-case local basis. Some global trends could emerge, however (EPA 2000):

- Warmer temperatures will probably increase evaporation and will thus probably increase precipitation in many areas, particularly at high latitudes, where every 0.5°C (0.9°F) increase in global average temperature corresponds to about a 5% increase in precipitation.
- Warmer temperatures will likely raise sea level in two ways—first, by melting existing ice masses such as mountain glaciers, the Greenland ice sheet, and (later) Antarctic ice, and second, by thermal expansion of seawater—like most other substances, water expands when heated. Some sea level rise is inevitable; the question is how much will occur. According to Titus and Naranyan (1995), there is a 50% chance that the rise will exceed 50 cm by 2100.
- Warmer temperatures may increase the severity of weather, although this topic is subject to much debate. Heat provides the energy for tornadoes, hurricanes, thunderstorms, and other severe weather patterns, so this is not inconceivable. There has been an increase in the frequency of severe high-rainfall storms over the past century as warming has commenced.

There is one more possible outcome from global warming, and it is a potentially major one. There is a great deal of methane trapped in sediments along continental margins as **methane gas hydrate**, a sort of water-methane ice material. The size of this reservoir of methane is difficult to ascertain because of its location, but estimates range from 6,000 to 15,000 GtC (likely at least double our remaining fossil fuel reserves, for reference). This gas hydrate is currently frozen, but it is close to its melting point. It is possible that with just a few degrees of warming, most or all of this material could melt, releasing a giant blast of methane into the ocean and then into the atmosphere.

This sounds outlandish, but geologic history tells us it is not. The Paleocene epoch, which led off the Tertiary period, was one of the warmest times in the last 65 million years (figure 6). The warmest part of this time occurred in a burst near the end called the Latest Paleocene Thermal Maximum (LPTM), when the deep ocean warmed about 4 degrees Centigrade (7.2 degrees Fahrenheit) in less than 10,000 years (Kennett and Stott 1991). At the same time, the shells of organisms growing in that bottom water showed an isotopic shift toward much lower concentrations of ^{13}C, and methane ice is typically extremely depleted in ^{13}C. Some scientists think, therefore, that the warm temperatures of the Paleocene led to a melting of methane hydrates and subsequent release of trillions of tons of methane to the ocean (Dickens et al. 1995). If we cause enough warming now, we might also cause the methane to destabilize, possibly doubling or tripling whatever warming we cause through fossil fuels. Many current models and studies indicate a LPTM-scale methane event is unlikely in the near future, but given what we know of the Paleocene, it is definitely worthy of study and consideration.

Conclusion

Global warming is almost undoubtedly a reality, and not a future one but a present one. Global temperatures have increased to 0.3°C (0.5°F) this century, well past their range of variability over the past 600 years. Atmospheric CO_2, a known greenhouse gas, is the likely culprit, having also increased by at least 25% this century. Climate proxy data sets show us that the potential warming we face is unknown in historic time and exceeds even the warmest parts of the last 150,000 years. However, there have been warmer times in the earth's past.

Climate models and flux diagrams help us to understand the climate dynamics and to analyze possible changes we can make. The biggest cause of global warming is the use of fossil fuels, and if that remains unchecked, we face warming, climate anomalies, sea level rise, and possibly more severe weather. These results may be mitigated or (more likely) amplified by climate feedback. However, we also may gain large tracts of more useful land in northern Asia and Canada. It is likely that global warming is something we want to prevent (all of us except Siberians, anyway), but we may not be willing to pay the price in energy costs and economic upheaval. A decision not to address the issue, or even a lack of a decision, will only let us put off until later dealing with our addiction to fossil fuels.

REFERENCES

Barnola, J. M., D. Raynaud, and Y. S. Korotkevich. 1987. Vostok ice core provides 160,000-year record of atmospheric CO_2. *Nature* 329:408–14.

Chappellaz, J., and J. Jouzel. 1992. Vostok Ice Core Data Set. IGBP PAGES/World Data Center-A for Paleoclimatology Data Contribution Series # 92-018. Boulder, Colo.: NOAA/NGDC Paleoclimatology Program.

Dickens, G. R., J. R. O'Neil, D. K. Rea, and R. M. Owen. 1995. Dissociation of oceanic methane hydrate as a cause of the carbon isotope excursion at the end of the Paleocene. *Paleoceanography* 10(6):965–72.

Eddy, J. A., and R. S. Bradley. 1996. Changes in time in the temperature of the earth. In *The Earth System Science Community: Curriculum,* edited by M. Keeler. http://www.circles.org/Round3/Curric/curric1.html, visited 1/20/2000.

Intergovernmental Panel on Climate Change (IPCC). 1995. *IPCC second assessment—Climate change 1995: A report of the intergovernmental panel on climate change.* Geneva, Switzerland: IPCC.

Jouzel, J., N. I. Barkov, J. M. Barnola, M. Bender, J. Chappellaz, C. Genthon, V. M. Kotlyakov, V. Lipenkov, C. Lorius, J. R. Petit, D. Raynaud, G. Raisbeck, C. Ritz, T. Sowers, M. Stievenard, F. Yiou, and P. Yiou. 1993. Extending the Vostok ice-core record of paleoclimate to the penultimate glacial period. *Nature* 364:407–12.

Keeling, C. D., and T. P. Whorf. 1999. Atmospheric CO_2 records from sites in the SIO air sampling network. In *Trends: A compendium of data on global change.* Oak Ridge, Tenn.: Carbon Dioxide Information Analysis Center, Oak Ridge National Laboratory, U.S. Department of Energy.

Kennett, J. P., and L. D. Stott. 1991. Abrupt deep-sea warming, palaeoceanographic changes and benthic extinctions at the end of the Palaeocene.
Nature 353:225–28.

Mann, M. E., R. S. Bradley, and M. K. Hughes. 1998. Global-scale temperature
patterns and climate forcing over the past six centuries. *Nature*
392:779–87.

Mitchell, C., and B. Geerts. 1997. Greenhouse warming: Facts and doubts. In
Climates and weather explained: The web site, edited by E. Linacre and B.
Geerts. http://rsd.gsfc.nasa.gov/912/geerts/cwx/notes/chap15/
clim_models.html, visited 1/20/2000.

Petit, J. R., J. Jouzel, D. Raynaud, N. I. Barkov, J. M. Barnola, I. Basile, M. Benders,
J. Chappellaz, M. Davis, G. Delaygue, M. Delmotte, V. M. Kotlyakov, M.
Legrand, V. Y. Lipenkov, C. Lorius, L. Pépin, C. Ritz, E. Saltzman, and M.
Stievenard. 1999. Climate and atmospheric history of the past 420,000
years from the Vostok ice core, Antarctica. *Nature* 399:429–36.

Post, W. M., T. Peng, W. R. Emanuel, A. W. King, V. H. Dale, and D. L. DeAngelis.
1990. The global carbon cycle. *American Scientist* 78:310–26.

Titus, J. G., and V. Narayanan. 1995. *The probability of sea level rise.* Washington,
D.C.: U.S. Environmental Protection Agency. EPA 230-R95-008.

U.S. Environmental Protection Agency (EPA). 2000. *The EPA global warming site,*
http://www.epa.gov/globalwarming/index.html, visited 1/22/2000.

Wilson, A. 1993. Cement and concrete: Environmental considerations.
Environmental Building News 2(2).
http://www.buildinggreen.com/features/cem/cementconc.html.

SUGGESTED READINGS

GENERAL

Intergovernmental Panel on Climate Change (IPCC). 1995. IPCC second
assessment—Climate change 1995: A report of the intergovernmental
panel on climate change. Geneva, Switzerland: IPCC. (Third assessment
due out soon.)

Kalkstein, L. S., B. M. Yarnal, and J. D. Scheraga, eds. 1998. Regional assessments of
climate change and policy implications. *Climate Research* 11(1).

Philander, S. G. 1998. *Is the temperature rising?: The uncertain science of global
warming.* Princeton, N.J.: Princeton University Press.

Watson, R. T., M. C. Zinyowera, and R. H. Moss, eds. 1998. The regional impacts of climate change: An assessment of vulnerability. Geneva Switzerland: IPCC.

POTENTIAL SEA LEVEL RISE

Titus, J. G., and V. Narayanan. 1995. The probability of sea level rise. Washington, D.C.: U.S. Environmental Protection Agency. EPA 230-R95-008.

HISTORY (AND PREHISTORY) OF CLIMATE CHANGE

Barnola, J. M., D. Raynaud, and Y. S. Korotkevich. 1987. Vostok ice core provides 160,000-year record of atmospheric CO2. *Nature* 329:408-14.

Kennett, J. P., and L. D. Stott. 1991. Abrupt deep-sea warming, palaeoceanographic changes and benthic extinctions at the end of the Palaeocene. *Nature* 353:225-28.

Mann, M. E., R. S. Bradley, and M. K. Hughes. 1998. Global-scale temperature patterns and climate forcing over the past six centuries. *Nature* 392:779-87.

National Geophysical Data Center—Paleoclimatology Program, http://www.ngdc.noaa.gov/paleo/paleo.html.

GLOBAL WARMING

American Meteorological Society, http://www.ametsoc.org/AMS.

Carbon Dioxide Information Analysis Center, http://cdiac.esd.ornl.gov.

Center for International Climate and Environmental Research—Oslo, http://www.cicero.uio.no/index_e.asp.

Center for International Earth Science Information Network (Columbia University), http://www.ciesin.org.

Global Change Data Center (NASA), http://www-tsdis.gsfc.nasa.gov/gcdc/gcdc.html.

Global Change Master Directory (NASA), http://gcmd.gsfc.nasa.gov.

Hadley Center for Climate Prediction and Research, http://www.met-office.gov.uk/research/hadleycentre/index.html.

U.S. Environmental Protection Agency. *The EPA global warming site*, http://www.epa.gov/globalwarming/index.html.

CHAPTER

2

Beyond the Horizon

UNDERSTANDING
THE ATMOSPHERIC
PHYSICS OF
CLIMATE CHANGE

Richard B. Kay

and

Jonathan A. R. Rall

Two trends have emerged during the latter part of the twentieth century that have caused scientists to become increasingly concerned about the health of the earth's atmosphere and, more generally, the integrated earth system. The earth system includes the atmosphere, oceans, **lithosphere** (solid earth), **cryosphere** (frozen ice caps, e.g., Antarctica and Greenland), **biosphere**, and the complex interactions among these several component systems. The first trend is a significant reduction in **stratospheric ozone** (O_3), a protective layer that shields the earth's surface from harmful radiation, over Antarctica that was first reported in 1985 (Farman et al. 1985). This **ozone depletion** was found to occur rapidly during the austral, or Southern Hemisphere, spring, resulting in an **ozone hole** that has increased steadily in size since it was first identified in 1970. Figure 1 displays recent data from the NASA Total Ozone Mapping Spectrometer (TOMS), flown most recently on an Earth Probes satellite.

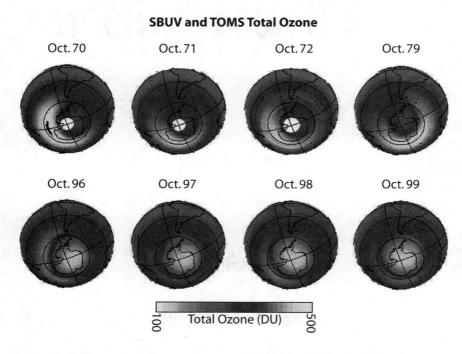

Figure 1. Eight years of ozone measurements over Antarctica depicting the ozone hole. Early data from Solar Backscatter Ultraviolet (SBUV) Spectrometer. More recent data from the Total Ozone Mapping Spectrometer (TOMS), NASA Goddard Space Flight Center. (See http://jwock.gsfc.nasa.gov and click Ozone.) A Dobson Unit (DU) is a measure of the total amount of ozone contained in a column (tube) at the place of interest.

The second trend is the slow rise of global average temperatures, about 0.7°C (1.3°F) since 1860. The temperature rise begins near the end of the **Industrial Revolution** (1830–1865) and correlates with ever increasing use of fossil fuels. The increase between 1980 and the present is particularly large, as can be seen in figure 2. The average global temperatures are

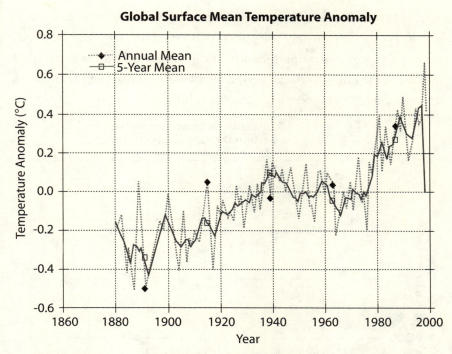

Figure 2. Global average temperature: 1860 to 2000. Note the strong increase from 1980 to 1998. The reference temperature of 0°C is set to the average from 1961 to 1990. Temperature anomaly in this case refers to temperatures above or below the 1961 to 1990 average. See http://www.giss.nasa.gov/data. See also Jones et al. (1999).

higher than they have been since 1400 A.D., the beginning of the Little Ice Age (Mann et al. 1998). Direct measurements of temperature have been available since around 1880, and indirect estimates (**climate proxies**) of ancient surface temperature are made from isotopic ratios of atmospheric constituents contained in air bubbles entrained in glacial ice (see Dobson, this volume). An **isotopic ratio** is a ratio of atoms/molecules with the same number of electrons but different nuclear composition. Certain of these nuclear components change with time, providing a time marker to the scientist. Additionally, atmospheric concentrations of CO_2, an important greenhouse gas, have increased from a pre–Industrial Revolution level of approximately 275 ppmv (parts per million by volume) to the current level of approximately 360 ppmv (figure 3). This is the highest atmospheric CO_2 concentration measured or documented on the earth since the end of the last interglacial period about 140,000 years ago (see Vostok ice cores data in figure 5 in Dobson, this volume).

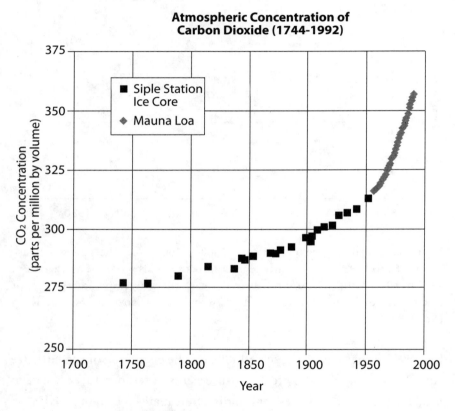

Figure 3. Carbon dioxide concentrations from remote global sites: 1744 to 1992. Siple Station (Antarctica) ice core data from Neftel et al. (1994); Mauna Loa data from Keeling and Whorf (1994).

Is CO_2 the culprit it has been made out to be or are other, more guilty suspects out there influencing the climate of our planet? To answer this question and others about our climate, we must try to understand our atmosphere and the various interworkings going on daily within it.

Obtaining such understanding is a daunting task as complex interactions take place between certain of the atmospheric gases, and between the atmosphere, earth, and oceans. And all of these interact with solar radiation, which is the energy source powering the earth.

The earth's atmosphere, while appearing at times perfectly blue, infinite, and simple, is in fact very complex. The first indicators of complexity are clouds: their formation, varied shapes, particle composition, size and morphology, the many different altitudes at which they form, and their motions. Clouds form when water vapor condenses or deposits on nucleating particles in the atmosphere. Nucleating particles then accumulate moisture (**accretes**) to form the cloud. In the process of forming a cloud, energy is exchanged (**latent heat of condensation/fusion**). Heat is given up to the atmosphere when water condenses or freezes. This is the primary mechanism by which energy is exchanged between the surface of the earth (evaporation, evapotranspiration) and deposited in the atmosphere (condensation).

The next indicator of complexity is the lateral, or horizontal, transport of water vapor throughout the atmosphere, referred to as **advection**. Water vapor is advected by winds that are generated by localized thermal heating, the earth's rotational motion, and gravity. Water vapor is arguably the most important trace species in our atmosphere; it is the primary greenhouse gas, having strong absorption features from just beyond the visible spectrum to the infrared and longer portion of the electromagnetic spectrum (see the section on the sun, this chapter). Water vapor provides the mechanism by which heat/energy is transported throughout the atmosphere.

The atmosphere carries not only water vapor but also many other trace species, aerosols, and dust. Some of the trace species have a strong effect on global warming, which we will discuss in some detail later. Others, like the **chlorofluorocarbons** (**CFCs**, complex molecules containing primarily chlorine, fluorine, and carbon atoms) in addition to being good infrared (IR) absorbers, are involved in the destruction of **ozone** in the stratosphere. The primary driving force for all of the interactions is the solar energy received by our planet, how it is absorbed, distributed, reflected, and reradiated.

Scientific Concepts

THE SOLAR RADIATION BUDGET

To understand how the globe maintains its more or less stable average temperature of about 15°C (59°F), we must look at the energy balance between solar input and the global output (figure 4). The earth is heated by solar energy, some of which is absorbed directly by atmospheric gases and clouds, and some that penetrates the atmosphere to be absorbed by land surfaces, vegetation, and bodies of water.

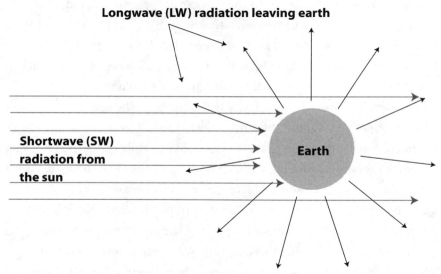

Figure 4. Incoming SW radiation from the sun impinges on the earth and its atmosphere. The heated earth-atmosphere system emits LW radiation in all directions.

Radiation from the sun is composed of different frequencies of electromagnetic radiation. These frequencies (f) are measured by their wavelengths (λ), and their relationship is $\lambda = c/f$, in which c is the speed of light. Our eye is sensitive to the wavelength region between 400 nm (violet) and 700 nm (red); nm is an abbreviation for nanometer, which is a billionth of a meter or 10^{-9} meters. This region between 400 and 700 nm is called the **visible region**, and it surrounds the peak of the solar spectrum, which is yellow, or the color that the sun appears. However, there is important radiation below the visible region called **ultraviolet** (**UV**) **radiation**, and above the visible region, from 800 nm to 5 μm, called **near infrared** (**NIR**) **radiation**. (Infrared wavelengths are often expressed in μm, or 10^{-6} m. There are 1,000 nm in 1 μm, so 2,000 nm equals 2 μm, or as often said in short, 2 microns.) Beyond the NIR is the mid-IR from 5 μm to about 30 μm and the far-IR from 30 μm to more than 200 μm. Beyond about 1.1 μm, infrared emission is primarily heat or thermal radiation; for example, the heat we feel on our hand from an incandescent light source.

For this discussion, solar radiation is called **shortwave** (**SW**) **radiation**, and is composed of visible, UV, and NIR radiation. The average power arriving per unit area at our planet's average distance from the sun is known as the **solar constant**. This quantity is called the solar irradiance, I_{sun}, which equals 1,368 joules/(second-meter²). A joule/sec (1J/s) is a unit of power that equals 1 Watt (1W). Our planet has a cross-sectional area $A = \pi r^2$, in which the radius of the earth, r, equals 6.37×10^6 m.

equation 1: Power $= I_{sun}A = I_{sun}\pi r^2 = 1.76 \times 10^{14}$ kW

So the earth intercepts a power, $P_{in} = 1.76 \times 10^{14}$ kW. In one hour we receive from the sun approximately 20 times as much energy as all U.S. electrical power plants produce in one year.

As the earth rotates, the solar radiation becomes averaged over the spherical surface during the course of a day. To calculate the earth's energy budget, we assume a uniform solar input over the surface area of the globe, $4\pi r^2$. This average input is called the solar flux, F_{sun}. Equating the average solar flux times the surface area of the globe to the solar irradiance intercepted by the earth's surface area (equation 1) gives:

equation 2: $(F_{SUN})(4\pi r^2) = I_{sun}\pi r^2$

Solving for F_{SUN} yields $F_{SUN} = I_{sun}/4 = 342$ W/m². This is the input solar flux to the earth.

The warmed earth acts much like a hot radiator in your home and emits infrared or **longwave (LW) radiation**. The wavelength of this earth radiation is centered at 10 μm with the majority of the emission occurring between 8 and 20 μm. The Stefan-Boltzmann Law can be used to calculate the power per unit area emitted by the earth. This equation represents the earth as a **blackbody radiator**:

$$\text{equation 3:} \quad F_{EARTH} = \sigma T_e^4$$

In this equation, $\sigma = 5.67 \times 10^{-8}$ W/(m² K⁴) is a constant and T_e is the temperature of the earth in Kelvin. To be in equilibrium:

Shortwave (SW) radiation from the sun = Longwave (LW) radiation leaving the earth. So

$$\text{equation 4:} \quad F_{SUN} (1-\alpha) = F_{EARTH} = \sigma T_e^4$$

The symbol α is the **albedo**, which is the fraction of the solar radiation reflected by the earth back into space (hence $[1-\alpha]$ is the amount transmitted into the earth). Equation 4 can be solved to obtain the average temperature of the earth:

$$\text{equation 5:} \quad T_e = [F_{SUN} (1-\alpha)/\sigma]^{1/4}$$

With the value of α at approximately 0.31, equation 5 projects a value of 254 K, or −17°C (1.4°F), for the earth's surface. This is the temperature of the earth as would be inferred from space. In fact, the earth's surface temperature is about 288 K (+15°C). The difference is due to the greenhouse effect.

Figure 5 depicts the global mean energy budget. Of the incident solar flux of 342 W/m², we see that 67 W/m² is absorbed by clouds, water vapor (H_2O), ozone (O_3), and aerosols. Note that the total reflected shortwave flux is 107 W/m² or 31% of 342 W/m² reflected back into space. This is

Beyond the Horizon

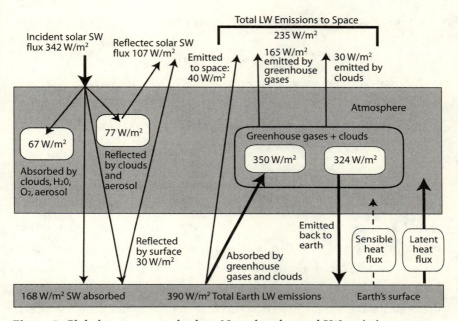

Figure 5. Global mean energy budget. Note that the total LW emissions to space are balanced by the net solar SW flux (342 W/m² minus 107 W/m² = 235 W/m²). Recent satellite data help in determining the balance. Figure based on data from Kiehl and Trenberth (1997).

what we mean by the albedo equaling 0.31. Note that clouds play a dual role, being involved in both the reflection and the absorption of solar radiation. Anything that would change the average amount of cloud cover would affect the energy input to the earth.

Having looked at the incoming solar radiation, now consider the LW radiation coming from the earth and clouds. On the right hand side of figure 5 are the LW radiation components. According to the Stefan-Boltzmann Law (equation 3), the surface of the earth emits approximately 390 W/m² when at 288 K. Note that only 40 W/m² is emitted directly to space. Of the 342 W/m² absorbed into the atmosphere, approximately two-thirds is absorbed by water vapor, CO_2, and aerosol with the rest being absorbed by clouds. These in turn radiate 195 W/m² to space and 324 W/m² back to the surface of the earth. It is important to note that in this case, cloud emissions contribute 30 W/m² to the LW outgoing radiation.

Looking at this result from the standpoint of the earth's surface, we have:

Solar radiation into surface	168 W/m^2
− LW from warm surface	− 390 W/m^2
+ LW in from clouds, aerosols, CO_2, H_2O	+ 324 W/m^2
Total	+ 102 W/m^2

The earth's surface is receiving 102 W/m^2 more radiative energy than it is emitting. This is referred to as **forcing**. As a result, the earth's surface would warm more were it not for evaporation from the oceans and thermal conduction of heat energy between the atmosphere and the earth's surface. Conduction is the **sensible heat flux** and evaporation is the **latent heat flux** shown at the right in figure 5. Together these total (−)102 W/m^2, the negative value indicating that this flux is leaving the earth's surface. In this case, the total radiative forcing to the earth's surface is balanced exactly by conduction and evaporation.

From the global mean energy budget, most of the important players involved in climate and global warming are identified. Water vapor, CO_2, and O_3 are the most active players in the greenhouse effect. In a greenhouse the glass is transparent to incoming solar radiation (SW) but opaque to thermal or heat radiation (LW), thereby trapping the heat within the greenhouse and keeping it warm. Similarly, the earth's atmosphere is transparent to visible (SW) radiation but much less transparent to LW radiation, in effect trapping some heat. Evaporation (and transportation) of water vapor is also very important. But the most important player in the overall energy budget is the sun itself.

The Sun

The sun is a gaseous ball, composed primarily of hydrogen nuclei undergoing nuclear fusion reactions as a result of being heated by gravitational attraction. Because its mass and size are modest, it is considered a normal or typical star. After going through its organizational phase some 4.5 billion years ago, it settled down for a 10-billion-year lifespan to convert hydrogen into helium and other heavier elements. Astrophysicists concur that our sun is relatively stable with the pressure from the hot fusion gases in the core balanced by the inward gravitational pull of the outer shells. The sun's output slowly increases over time, having increased by approximately 30% over the last 4 billion years. This rate of increase is very slow, on the order of seven parts per billion per year, due to burning of helium and other heavy element products. So, on a time scale of a hundred or thousand years, this increase can be considered negligible.

However, evidence is accumulating that suggests the solar output may change by 1% or more over relatively short times, on the order of a few hundred years. There are records of periodic sunspot activity going back to the time of Galileo, who first showed that sunspots move across the face of the sun and used them to determine the sun's twenty-five-day rotation period. In a careful study, John Eddy (1977) compiled historical and more modern data to produce the plot of sunspot number versus year, displayed in figure 6. The maxima repeat with a cycle of some nine to thirteen years, with an average cycle being about eleven years long. Annual maxima range from as few as 20 to as many as 180 sunspots during a peak period called the **solar maximum** (the **solar minimum** refers

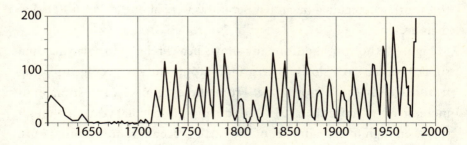

Figure 6. Sunspot activity from 1600 to present. Note the eleven-year cycle in solar maxima, which are the regions of maximum sunspot activity. Data from Eddy (1977).

to the lowest level of annual sunspot activity). Sunspots indicate strong magnetic activity and are surrounded by regions called **faculae**, bright regions of intense activity (see Lean et al. 1995). Eddy also found a strong correlation between sunspot activity and the Australis and Aurora Borealis. The aurora phenomenon is due to high-speed charged particles in the "solar wind" being trapped in the earth's magnetic field and caused to spiral toward the poles. These high-energy particles strike and excite atmospheric gases on the way, creating beautiful light emissions, seen in the higher latitudes in both hemispheres. This result implies increased solar activity during periods of frequent sunspots.

An analysis of recent satellite data shows that the number of sunspots is positively correlated with solar output, particularly in the ultraviolet. Using several sets of recent satellite data, Crommelynck and Dewitte (1997) found a change of about 2.3 W/m^2 in solar radiance between the solar maximum and minimum between 1979 and 1987 and about 1.4 W/m^2 between the 1987 minimum and the 1992 maximum. Output was highest when the sunspot count was greatest and lowest during the sunspot minima in 1986. These minima and maxima can also be seen in figure 7 (top) where the maxima around 1980 and 1990 are easily visible. Figure 7 (bottom) shows the reconstructed solar irradiance from the 1500s to the present. Note in figure 6 that there was very little sunspot activity between 1600 and 1720, a time called the **Maunder minima**. This time coincides in part with the Little Ice Age, shown in figure 7 (bottom), when global mean temperatures were about 1°C (1.8°F) lower than present.

Figure 7 (bottom) additionally shows peaks related to solar maxima superimposed on a slowly increasing output, and the averaged forcing irradiance on the left ordinate. From these data we can estimate an increase of no more than 0.6 W/m^2 during the last 500 years. Although this is enough to explain part of the earth's recent gradual temperature increase, it cannot account for the higher rates of increase seen during the last thirty years. Here we must invoke the effects of the atmosphere and increased greenhouse gas emissions.

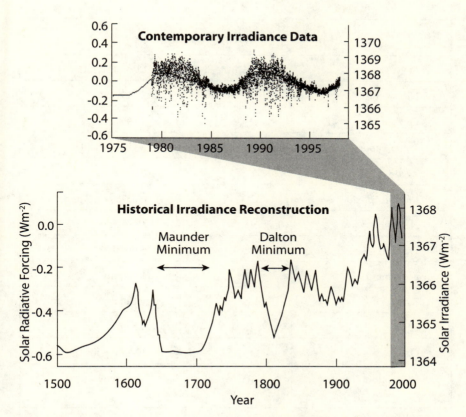

Figure 7. Solar output 1500 to present. Constructed from satellite data from 1980 to present (top) and from prior to 1980 (bottom). Reprinted by permission from Lean et al. (1995) and Rind, Lean, and Healy (1999).

WATER VAPOR AND GLOBAL CIRCULATION

Next to the primary atmospheric components N_2 and O_2, water vapor is the most prevalent species in our atmosphere (table 1). The highest global concentrations of water vapor are found in the equatorial regions, where mixing ratios as high as 20 grams of water vapor per kg of dry air (about 2.8% by weight) are found (figure 8). The mean water vapor density decreases with increasing ± latitude. It also decreases exponentially with increasing altitude, being mostly confined to the first one to two km of the atmosphere above the earth's surface.

TABLE 1. Atmospheric Constituents		
CONSTITUENT	SYMBOL	CONCENTRATION
Nitrogen	N_2	78.1%
Oxygen	O_2	20.9%
Water Vapor	H_2O	highly variable
Argon	A	0.93%
Carbon Dioxide	CO_2	0.037% (370 ppmv)
Neon	Ne	18 ppmv
Helium	He	5 ppmv
Methane	CH_4	1.7 ppmv
Krypton	Kr	1.1 ppmv
Hydrogen	H_2	0.5 ppmv
Ozone	O_3	10–100 ppbv
Chlorofluorocarbons	CFC-11+ CFC-12	~0.5 ppbv

Quantities in percent, parts per million by volume (ppmv), or parts per billion by volume (ppbv)

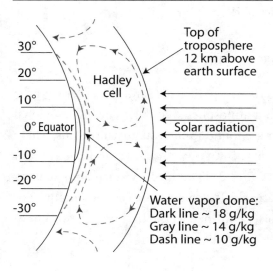

Figure 8. Tropical Hadley cells and water vapor distribution near the equator. Figure is for equinox solar position. Heights of water vapor contours exaggerated approximately a factor of 2 for visibility. Water vapor data from Oort and Peixoto (1983).

A region of intense thunderstorm activity near the equator is called the **Intertropical Convergence Zone (ITCZ)**. This organized activity moves north of the equator in the Northern Hemisphere summer and south of it in winter. The ITCZ closely matches the region where solar radiation strikes the globe at nearly right angles, is minimally absorbed by the atmosphere, and heats the tropical oceans and land areas. Solar radiation is more strongly absorbed here than at higher latitudes, whereas the LW earth emissions are more uniformly distributed over the planet. The LW emissions are actually greater at the poles than solar SW heating. Consequently, energy absorbed in the equatorial regions moves toward the cooler poles. Approximately half of the heat energy is moved from the tropics by the atmosphere and the other half by ocean currents.

Heated air rises near the ITCZ, setting up a circulation called a **Hadley cell** (figure 8). After rising, the air spreads north and south in a pumping cycle that returns earthward after cooling, near the latitudes ±30°. The return circulation picks up moisture from subtropical oceans on its way toward the equator, creating convective thunderstorm activity with heavy rains in the ITCZ. The Hadley cell conveyor belt carries water vapor upward and toward the north and south poles where it is cooled and condensed into clouds, releasing large amounts of energy that heat the atmosphere (see Graedel and Crutzen 1993).

As energy absorbed in equatorial regions is started on its way toward the poles, the trip involves uneven heating and cooling and air motions strongly influenced by the earth's rotation. For example, in the Northern Hemisphere, the air at the bottom of a Hadley cell moves southward while the earth turns eastwardly under it, creating a strong wind that travels toward the southwest, but is named according to its source direction, the northeast. As this wind approaches the equator, it couples with the lower part of a clockwise circulation that forms over all Northern Hemisphere oceans. Known as the Northeast Trade Wind, it blows at an average of 20 knots. The Northeast Trade Winds move with the seasons, in a band centered around latitude 18°N in the summer and around 10°N in the winter. A similar circulation sets up in the Southern Hemisphere, where the circulation is counterclockwise over the oceans. The circulation picks up water vapor from evaporation, elevates it, and condenses it into clouds from which it eventually leaves as precipitation.

CLOUDS

The net effect of clouds on the earth's radiation budget is not conclusively determined. Low-altitude clouds (e.g., cumulus, stratus) act to cool the earth by reflecting SW radiation and emitting some LW radiation to space (figure 5). Higher-altitude clouds such as cirrus are transparent to incoming SW radiation but reflect outgoing LW radiation back toward the surface, thereby acting to heat the earth. Therefore, increasing low-altitude cloudiness would have a net cooling effect whereas increased cirrus cloud formation would tend to have a net heating effect on the earth. It is clear that clouds are one of the greatest sources of uncertainty in predicting climate change. All clouds, regardless of altitude, are made of water droplets and/or ice crystals and are the strongest absorbers of LW radiation in the atmosphere. Their water droplet/ice crystal size distribution varies among cloud types, which affects the way they reflect, absorb, and transmit SW and LW radiation. **Condensation nuclei** are essential for cloud formation with wind-blown dust, sea salt, pollen, volcanic debris, and algae acting as naturally occurring cloud-condensing nuclei (CCN). Anthropogenically produced (caused by humans) aerosols, which are stable groupings of solid or liquid particles in the air, are increasingly important in cloud formation. Aerosols may be formed from combustion products, dust and emissions from construction and agricultural activities, industrial emissions, and soot from biomass burning, among others. It turns out that one of the most important condensation nuclei over ocean areas is sulfate aerosols formed from dimethylsulfide emitted by **phytoplankton,** which is composed of the minute plant life living in the water column (Charlson et al. 1987). This aerosol would likely be produced more rapidly with increased global warming, thus creating more cloud formation over the warmed oceans.

THE EARTH RADIATION BUDGET EXPERIMENT

In the 1980s, NASA began a program called the Earth Radiation Budget Experiment (ERBE). Satellites launched for this experiment in the mid-1980s monitored both LW and SW radiation. In July 1985, it was found that clouds provide a net forcing of about (–)17 W/m^2 (Ramanathan et al. 1989). The global average cloud-forcing components were:

$$
\begin{aligned}
\text{LW forcing} \ &= \ +30.1 \ \text{W/m}^2 \ \text{(greenhouse warming)} \\
\text{SW forcing} \ &= \ -46.7 \ \text{W/m2} \ \text{(cloud solar reflection)} \\
\hline
\text{Net forcing} \ \ &\ \ \ -16.7 \ \text{W/m}^2
\end{aligned}
$$

These data supported atmospheric climate modeling results, in particular by **general circulation models** (**GCMs**), which had been predicting such an effect (see Dobson, this volume). The GCMs results suggest that clouds provide a net cooling of the earth and partially counter the greenhouse effect.

One possible scenario has a forcing mechanism warming the ocean surface, leading to increased dimethylsulfide production by plankton and increased surface evaporation. This, in turn, leads to increased sulfate aerosols in the troposphere and hence increased cloud production, providing a negative feedback mechanism to help contain the warming. A recent study of SW radiation over the warm pool in the Pacific (Ramanathan et al. 1995) indicated that clouds over the pool reflected an increased amount of SW radiation back to space. The clouds also trapped more SW radiation than predicted by current models. Globally, this phenomenon would imply a change in the radiation budget shown in figure 5, reducing the overall average energy absorbed by the earth by up to 25 W/m^2 (Cess et al. 1995). Several studies have found that cloud cover over landmasses seems to increase during warm periods, relative to that during cold periods. Observations made during the last century in a regional study of the United States from 1900 to 1987 found the percentage of cloud cover over coastal southwest, coastal northeast, and the southern plains positively correlated with the increase in global mean surface temperature (Croke et al. 1999). Equally important, they found a lessening of the Azores high, the North Pacific high, and the Icelandic low with increasing global temperatures. This implies a possible link between cloud cover and weakening of long-standing pressure features. Just how this translates into climate change is not known at this time.

ROLES OF CO_2 AND O_3

Most molecules absorb and emit IR radiation, which is often called IR activity. The wavelengths of IR light that are emitted or absorbed are related to energy jumps between internal energy levels of the molecule and are unique to a particular molecule. We can identify such a molecule and estimate its abundance from these absorptions and emissions, usually referred to as its **IR spectrum**. The infrared activity of the important atmospheric species is shown in figure 9. Strong absorption bands of CO_2 between 13 and 17 μm and of O_3 between 9 and 10 μm are shown, along with a lesser absorption peak of methane (CH_4) between 7 and 8 μm. These are superimposed on water vapor (H_2O) absorptions covering the width of the plot. When we compare the increase in atmospheric CO_2 with the increase in global mean temperature, it is easy to conclude that further CO_2 increase may lead to further increase in global temperatures.

A study of fifteen GCMs by Cess et al. (1993) predicted temperature increases ranging from 1.9° to 5.3°C (3.4° to 9.5°F) with an average 3.8°C (6.8°F) and a doubling of atmospheric CO_2 levels from 330 ppmv to 660 ppmv. In addition, the models predicted that such an increase would produce a forcing of about 4 W/m^2. This implies an increase of around 1°C (1.8°F) for each W/m^2 of forcing. Although not shown in figure 9, CFCs are active between 8 and 10 μm, and, although present in small concentrations, they have large IR absorptions and are becoming significant in the greenhouse effect. It is estimated that between 1975 and 1985, all the other trace gases (CH_4, CFCs, N_2O, O_3, and others) contributed about the same greenhouse forcing as that of CO_2 (Ramanthan et al. 1989). That amounts to about 0.5°C (0.9°F) every ten years for at least the next fifty years, even if we stop putting all of these gases into the atmosphere today.

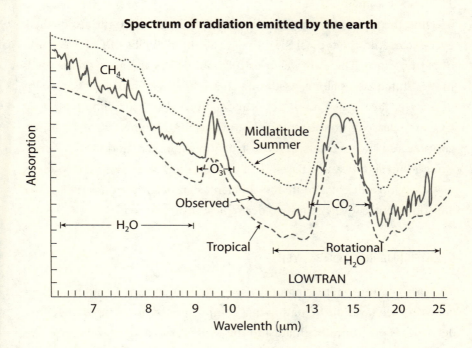

Figure 9. Atmospheric absorption (vertical axis) versus wavelength in microns. This is an inverted infrared absorption spectrum, showing some absorption peaks associated with the presence of certain atmospheric species. The solid curve is the measured spectrum with peaks associated with O_3 and CO_2, a small speak associated with CH_4, and two regions of H_2O absorptions identified. These absorptions are superimposed on the background continuum (blackbody). The dashed curves are from LOWTRAN (Low Resolution Transmission) calculations based on theoretical models for mid-latitude summer (fine dots) and for Tropical regions (larger dashes). Note the very large CO_2 peak, which is superimposed on the water vapor rotational absorption. From radiance data obtained over the Gulf of Mexico from NIMBUS 3 (Fenn et al. 1985).

Consider the following. When a gallon of gasoline is burned by combustion, in your car for example, it creates about 4 lbs of H_2O and 4 lbs of CO_2 as by-product. A typical car gets 25 miles per gallon and travels about 15,000 miles per year. So it produces:

P = (4 lbs/gallon) x (15,000 miles/25 mpg) = 2,400 lbs of CO_2 (per year)

Thus, the typical car emits nearly its weight in CO_2 into the atmosphere every year. And the typical SUV throws about 4,000 lbs of CO_2/year into our atmosphere. There are well over 100 million cars operating on the roads in the United States alone. Assuming an average CO_2 emission of 3,200 lbs of per year (average of cars + SUVs and light trucks), 320 billion pounds of CO_2 are released into the atmosphere per year. Add in heavy trucks, buses, gas-powered lawn mowers, and construction equipment and the CO_2 emissions go up significantly. Based on CO_2 emissions alone, we need to make new cars and trucks far more fuel efficient, significantly reduce their other emissions, and actively remove older vehicles from the road.

STRATOSPHERIC OZONE LAYER

The earth's atmosphere is naturally separated, or stratified, into many layers with each layer participating in a complicated, dynamic exchange that determines our overall climate and the earth's critical energy balance. We live in the **troposphere**, the thin layer that is adjacent to the earth's surface, and often fly through the **tropopause**, which is the region where the troposphere meets the overlying **stratosphere**; therefore, these are the two layers of which we are most aware (figure 10). The troposphere is subject to rapid heating and moisture exchanges with the surface. Temperature in the troposphere decreases with altitude, but warm air is more buoyant than cold air, so air warmed at the surface tends to move upward until it is neutrally buoyant or runs into something. Having colder air over warmer air, as in the troposphere, is an inherently unstable situation that can lead to significant weather phenomena. The troposphere is constrained by the overlying stratosphere. Oxygen (O_2) and ozone (O_3) play a very important role in the stratosphere, absorbing UV radiation (240–320 nm) and heating the stratosphere to create a temperature profile in which temperature increases with increasing altitude. Hence, the stratosphere is very stable with little vertical motion since warm air overlies cold air. The stratosphere caps the turbulent troposphere, and there is not much exchange between

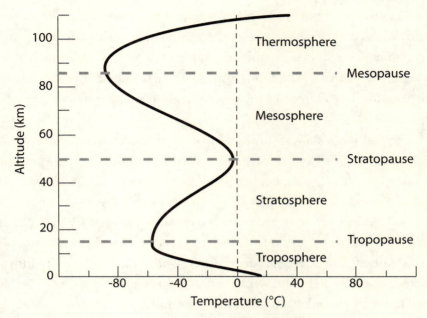

Figure 10. Atmospheric temperature as a function of altitude. We live in the troposphere, which extends from 0 to 12 km above the earth's surface. The stratospheric ozone layer is centered around 28 km above the earth's surface.

the two. One can even argue that the ozone and oxygen form the stratosphere by absorption of solar UV. With the discovery of the Antarctic ozone hole and the overall trend of decreasing global stratospheric ozone, there is less heating of the stratosphere by incident solar radiation. This could lead to a less well-defined tropopause and hence a less well-contained troposphere and ultimately changes in weather patterns as well as global climate.

The stratospheric ozone layer is a region of increased O_3 concentration extending from about 15 to 45 km above the earth's surface. This layer occurs naturally and is maintained by the absorption of energetic UV solar radiation by O_2 molecules, which dissociate into two oxygen atoms:

O_2 + UV photon $\Rightarrow$ O + O   (for light with wavelength ≤ 240 nm)
$O + O_2 \Rightarrow O_3$

The destruction of stratospheric ozone over Antarctica in the austral spring is a complicated sequence of atmospheric physics, chemistry, and dynamics that required extensive detective work to fully understand. Several unique factors, both human-made and naturally occurring, had to happen for the Antarctic ozone hole to develop and worsen over the past two decades.

The first was a slow, steady buildup over many decades of chlorofluorocarbons (CFCs) and chlorine reservoir compounds in the Antarctic stratosphere. CFCs were developed in the early 1930s by E.I. du Pont de Nemours and Company. Many CFCs, CFC-11 (R-11/Freon 11) and CFC-12 (R-12/Freon 12), have been widely used since the end of World War II to the present as refrigerants, blowing agents, and spray aerosol propellants. Although no longer produced in the United States or most industrialized countries, R-11 and R-12 are found throughout the United States and the world in older refrigerators, air conditioners, and other cooling equipment. One of the alleged benefits of these new refrigerants was that they were some of the most stable compounds ever created, with typical atmospheric lifetimes of 50 to 100 years. Although relatively heavy compared with molecules of N_2 and O_2, the major constituents composing air (table 1), their long lifetimes permit some fraction of the released CFCs to be lofted up into the stratosphere and then to migrate toward the poles. The stable CFC compounds are split apart by the action of high-energy UV sunlight, which is abundantly available in the stratosphere. They then form less stable reservoir chlorine compounds.

Second, during the austral winter, a highly stable circumpolar ocean current and wind (called a **circumpolar vortex**) forms around the perimeter of the Antarctic continent. This essentially isolates the polar atmosphere from the middle and lower latitudes so that there is little or no air exchange. The concentration of CFCs and chlorine reservoir compounds in the polar stratosphere has been steadily increasing over several decades, creating a kind of killing field for stratospheric ozone once the sun reappears in each spring.

Beyond the Horizon   53

Third, with an underlying landmass, the Antarctic continent, through radiative cooling, is significantly colder than the Arctic, which is frozen sea ice with underlying water. In the austral winter, the Antarctic stratosphere can cool to below 190 K. At these temperatures, clouds form in the lower stratosphere. A particular kind of **polar stratospheric cloud (PSC)** that forms is composed of frozen water and frozen nitric acid trihydrate (NAT). The importance of NAT is that it provides a surface that enables a particular chemical reaction to take place. In this reaction, high-energy solar UV liberates chlorine radicals, Cl-, from their reservoir compounds.

Once liberated the highly reactive Cl- radicals, coupled with energetic solar UV photons, then dissociate many O_3 molecules due to the catalytic nature of Cl in the reaction:

$$Cl + O_3 \Rightarrow ClO + O_2$$
$$ClO + O \Rightarrow Cl + O_2$$
$$\text{net: } O_3 + O \Rightarrow 2O_2$$

With the extensive formation of PSCs during the winter, the springtime ozone destruction process occurs very rapidly, resulting in an ozone hole that peaks in late September or early October as is seen in figure 11. The hole, which typically is at its maximum in mid to late October, does not fill in with ozone from middle latitudes because the circumpolar vortex does not break up until November or December. During this period the ozone hole (figure 1) is often oblong or erratically shaped, and precesses such that the area of thinnest ozone can reach to much lower latitudes, including the tip of South America. When the circumpolar vortex finally breaks up, ozone from lower latitudes rushes in to fill the void, resulting in a thinning of ozone over populated landmasses, including South Africa, Australia, New Zealand, and South America.

The overall trend of thinning ozone has resulted in increased surface UV-B radiation worldwide and has many far-reaching effects. UV-B

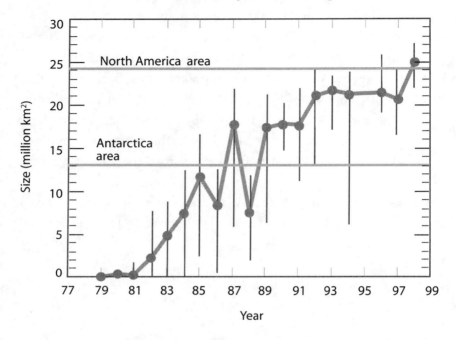

Figure 11. Size of ozone hole averaged from September 7 to October 13 (period of maximum springtime ozone depletion) for years 1977 to 1999. Hole size is defined by the 220 Dobson Unit contour. Mapped by the TOMS instruments identified in figure 1. For more information on ozone, go to: http://jwocky.gsfc.nasa.gov.

radiation (290–320 nm) is the most damaging to biological organisms and most efficiently blocked by the action of the stratospheric ozone. UV-B radiation causes agricultural changes, decreased plant biodiversity, increased oceanic salinity, and increased skin cancer rates.

Conclusion

Global temperatures have been on the rise since the beginning of the Industrial Revolution, but their increase has been more rapid particularly in the last twenty years. Anthropogenically produced and/or increased emissions, particularly CO_2, CFCs, and tropospheric or ground-level ozone, are greenhouse gases and are furthering to increase our average temperatures. The CFCs in particular have long atmospheric lifetimes and will take 50 to 100 years to be "cleansed" from the atmosphere by normal processes. The springtime ozone hole over both poles and the thinning ozone layer worldwide have many complex and far-reaching effects including changes in the structure of the atmosphere, in weather patterns, and in plant and animal biodiversity and increased skin cancers among humans.

The sun itself is in an increasing phase and is contributing to global warming but is likely not responsible for all of the current warming. Global climate models (GCMs) predict warming of 0.5°C (0.9°F) per decade for the foreseeable future. We have only warmed about 0.6°C (1.1°F) in the last 100 years, and half of that occurred between 1980 and the present. We also have ice core data that show that, following an ice age, when global temperatures are perhaps 6°C (10.8°F) lower than present, there is a temperature increase to near-current average global temperature. This increase takes place over a period of five to ten thousand years. After the ice age, 150,000 years ago, the warming proceeded to about 1.5°C (2.7°F) above the current average global temperature, then subsided. During the warming, the oceans effused stored CO_2, increasing atmospheric concentrations to about 200 ppmv. Why didn't this cause positive feedback and runaway temperatures due to an enhanced greenhouse effect? What stabilized the temperatures? There must be some compensating response or negative feedback in the system, such as increased low-level cloud cover, which allowed the earth to maintain its relatively stable global temperature over the millennia. However, the integrated earth system has never had to contend with a human population of more than 6 billion people. Increasing global population and increased standards of living result in increased per capita energy requirements and hence increased CO_2 and other greenhouse gas emissions. The human population has become a significant forcing function on global climate.

REFERENCES

Cess, R. D., M. H. Zhang, P. Minnis, L. Corsetti, E. G. Dutton, B. W. Forgan, D. P. Garber, W. L. Gates, J. J. Hack, E. F. Harrison, X. Jing, J. T. Kiehl, C. N. Long, J. J. Morcrette, G. L. Potter, V. Ramanathan, B. Subasilar, C. H. Whitlock, D. F. Young, and Y. Zhou. 1995. Absorption of solar radiation by clouds: Observations versus models. *Science* 267:496–99.

Charlson, R. J., J. E. Lovelock, and M. O. Andrae. 1987. Oceanic phytoplankton, atmospheric sulphur, cloud albedo and climate. *Nature* 326:655–61.

Croke, M. S., R. D. Cess, and S. Hameed. 1999. Regional cloud cover associated with global climate change: Case studies for three regions of the United States. *Journal of Climate* 12:2128.

Crommelynck, D., and S. Dewitte. 1997. Solar constant temporal and frequency characteristics. *Solar Physics* 173(1):177–91.

Eddy, J. A. 1977. The case of the missing sunspots. *Scientific American* 236:80–88.

Farman, J. C., B. G. Gardiner, and J. D. Shanklin. 1985. Large losses of total ozone in Antarctica reveal seasonal ClO_x/NO_x interaction. *Nature* 315:207–10.

Fenn, R. W., S. A. Clough, W. O. Gallery, R. E. Good, F. X. Kneizys, J. D. Mill, L. S. Rothmann, E. P. Shette, and F. E. Volz. 1985. Optical and infrared properties of the atmosphere (chapter 18). In *Handbook of Geophysics and the Space Environment*, edited by A. Jursa. Bedford, Mass.: Air Force Cambridge Research Laboratories.

Graedel, T. E., and P. J. Crutzen. 1993. *Atmospheric change: An earth system perspective.* New York: W. H. Freeman.

Jones, P. D., M. New, D. E., Parker, S. Martin, and J. G. Rigor. 1999. Surface air temperature and its change over the past 150 years. *Reviews of Geophysics* 37:173–99.

Keeling, C. D., and T. P. Whorf. 1994. Atmospheric carbon dioxide records from sites in the SIO air sampling network. 16–26. In *Trends '93: A compendium of data on global change*, edited by T. A. Boden, D. P. Kaiser, R. J. Sepanski, and F. W. Stoss. ORNL/CDIAC-65. Oak Ridge, Tenn.: Carbon Dioxide Information Analysis Center, Oak Ridge National Laboratory.

Kiehl, J. T., and K. E. Trenberth. 1997. Earth's annual global mean energy budget. *Bulletin of the American Meteorological Society* 78:197–208.

Lean, J., J. Beer, and R. Bradley. 1995. Reconstruction of solar irradiance since 1610: Implications for climate change. *Geophysical Research Letters* 22:3195.

Mann, M. E., R. S. Bradley, and M. K. Hughes. 1998. Global scale temperature patterns and climate forcing over the past six centuries. *Nature* 392:779–87.

Neftel, A., H. Friedli, E. Moore, H. Lotscher, H. Oeschger, U. Siegenthaler, and B. Stauffer. 1994. Historical carbon dioxide record from the Siple Station ice core (pp. 11–14). In *Trends '93: A compendium of data on global change*, edited by T. A. Boden, D. P. Kaiser, R. J. Sepanski, and F. W. Stoss. ORNL/CDIAC-65. Oak Ridge, Tenn.: Carbon Dioxide Information Analysis Center, Oak Ridge National Laboratory.

Oort, A., and J. Peixoto. 1983. Global angular momentum and energy balance requirements form observations. *Advances in Geophysics* 25:355–490.

Ramanathan, V., B. R. Barkstrom, and E. F. Harrison. 1989. Climate and the earth's radiation budget. *Physics Today* 22–32.

Ramanathan, V., G. L. Subasilar, G. J. Zhang, W. Conant, R. D. Cess, J. T. Kiehll, H. Grassi, and L. Shi. 1995. Warm pool heat budget and shortwave cloud forcing: A missing physics? *Science* 267:499.

Rind, D., J. Lean, and R. Healy. 1999. Simulated time-dependent climate response to solar radiative forcing since 1600. *Journal of Geophys. Res.* 104:1973–99.

SUGGESTED READINGS

Brasseur, G. P., and S. Solomon. 1986. *Aeronomy of the Middle Atmosphere,* 2nd edition. New York: Kluwer Academic Publishers.

Crutzen, P. J., and V. Ramanathan. 2000. Pathways of discovery: The ascent of atmospheric sciences. *Science* 290:299–304.

Eddy, J. A. 1977. The case of the missing sunspots. *Scientific American* 236:80–88.

Graedel, T. E., and P. J. Crutzen. 1993. *Atmospheric change: An earth system perspective.* New York: W. H. Freeman, chapters 3 and 4.

Ramanathan, V., B. R. Barkstrom, and E. F. Harrison. 1989. Climate and the earth's radiation budget. *Physics Today* 22–32.

Salby, M. L. 1996. *Fundamentals of atmospheric physics* 61, International Geophysics Series, edited by R. Dmoska and J. R. Holton. San Diego, Calif.: Academic Press, chapter 1.

FOR A CONTRARIAN POINT OF VIEW

Priem, H. N. A. 1997. CO_2 and climate: A geologist's view. *Space Science Review* 81:173–98.

CHAPTER 3

Ecosystems in Danger

AN ECOLOGICAL
PERSPECTIVE ON
CLIMATE CHANGE

Susan Herrod-Julius
and
John McCarty

Ecology, the scientific study of the relationship of organisms to their physical and biological environment, seeks to understand the factors that determine the **species distribution** and **species abundance** of plants and animals. This broad area of inquiry includes the study of how the behavior and the physiology of individuals affects their survival and reproduction, and how characteristics of individuals affect population dynamics. Ecology is also concerned with the interactions among populations that make up **ecological communities** and ultimately how all of these factors interact to influence the flow of energy and nutrients through **ecosystems.** Many fields that address pressing environmental problems rely on input from the science of ecology, including **conservation biology,** wildlife and fisheries management, and **environmental toxicology.**

Since its origin as a scientific discipline in the late 1800s, ecology has recognized climate as one of the most important determinants of **ecosystem structure** and **ecosystem function**. Because of the role climate plays in the abundance, distribution, and biology of species and ecosystems, ecologists seek to understand how species will be affected by changes in temperature and precipitation.

Recent changes in climate have been recorded through observations of surface temperature that show that there has been a global mean warming of 0.3° to 0.6°C (0.5° to 1.8°F) over the past century (Intergovernmental Panel on Climate Change [IPCC] 1996a). Atmospheric scientists have reached the consensus that rapid increases in the amount of carbon dioxide, or CO_2, in the atmosphere, resulting largely from the burning of fossil fuels, are contributing to the observed climate changes (IPCC 1996a). This impact of humans on climate raises the concern that climate change will continue and even accelerate in future years.

The response of species to ongoing changes in climate is complicated by the concurrent rapid rise in concentrations of CO_2 in the atmosphere. Plants use CO_2 in meeting their energy requirements through photosynthesis. Higher concentrations of CO_2 in the atmosphere may not have equal effects on the productivity of different plant species, leading to alterations in species compositions and ultimately to ecosystem-level changes.

Climate changes are occurring in the context of increased pressure from human activities (e.g., land use changes, introduction of nonnative species and diseases, and pollution). Ecologists are concerned that these pressures will lead to far-reaching effects on valued resources and species.

Just as humans can have impacts on ecosystems, so too can ecosystems have impacts on humans. Natural ecosystems produce valuable products used by our societies such as fish and wood products. The healthy functioning of ecosystems also provide less tangible services in both natural and human dominated ecosystems. These so-called **ecosystem services** are defined as being both conditions and processes that sustain and fulfill human life (Daily 1997). Ecosystem services maintain biodiversity, produce goods, and perform life-support functions. Without such services, human existence would be much more difficult. If changes in climate disrupt normal ecosystem functioning, vital processes such as purification of water and mitigation of floods and droughts could be impaired.

How species and ecosystems change as climate changes has implications beyond the impact on ecosystem services. Changes in ecosystem structure and function may also result in feedbacks to the climate system. Land surface changes such as those in the distribution of vegetation across the globe affect temperature and precipitation patterns. Large-scale changes in ecosystems may result in complex responses in atmospheric circulation. These interactions between the atmosphere and the biosphere could further accelerate climate change (IPCC 1996b).

Scientific Concepts

One of the primary goals of ecology is to understand how a multitude of factors in the environment interact to determine the distribution and abundance of plants and animals. These factors include the availability of food and other resources, interactions with predators, competitors, and parasites, as well as characteristics of the physical environment such as temperature, moisture, and availability of nutrients.

Ecologists target studies toward a number of different levels of ecological organization. The environment acts directly on individuals who must find the means to survive and reproduce. Characteristics of an individual's behavior and physiology determine what range of temperatures they can tolerate, how much water they need to survive, and what kinds of food they require. For individuals of a given species, the combination of environmental conditions where they can survive and reproduce is the species' **niche.** Within a given area, characteristics of individuals influence population-level characteristics such as population size and how fast population size changes.

Ecologists are also concerned about how populations of different species interact as predators, competitors, and parasites. Understanding interactions among these groups of species, or communities, is needed to understand characteristics of individuals and populations. Ecological communities in the larger context of their physical environment are called ecosystems. Characteristics of ecosystems frequently discussed include their structure and functioning. Ecosystem structure is the physical form of the ecosystem: the living and nonliving parts and how they fit together. The interactions of these parts are ecosystem functioning. Ecosystem functioning includes the exchange of water, nutrients, and energy among the abiotic and biotic parts of an ecosystem.

Humans are generally not thought of as being part of ecosystems. However, our dependence on ecosystem functioning and our ability to alter or manage ecosystems illustrate what an integral part of ecosystems we have become. Even systems that are heavily managed by humans, such as agricultural lands and forests, are governed by ecological processes that could be altered by climate change.

Humans depend on the goods and services provided by ecosystems such as seafood, timber, clean air, and clean water. The type and the amount of goods and services produced depend on ecosystem structure and functioning. For example, patterns of nutrient flow in healthy wetland ecosystems determine the ability of these communities to purify water. Threats to ecosystem structure and function, such as climate change, will result in losses to human economies that will be difficult and expensive to replace (Daily 1997). The challenge to ecologists is to try to understand how communities might change in order to best prepare for or prevent irreversible damage to natural and human-dominated ecosystems.

Approaches to Research in Ecology

Even though climate has long been known to be an important component of a species' niche, environmental factors interact in complex ways, making it difficult to determine how a change in climate will influence a given species. Understanding the possible implications of greenhouse-related climate

Ecosystems in Danger

warming is further complicated by the additions of CO_2 into the atmosphere. As well as being a potent greenhouse gas, CO_2 is a necessary resource used in **photosynthesis** by plants. Thus the challenge to ecologists is to understand how the addition of a resource (CO_2) and changes in temperature and precipitation will interact to change the distribution and abundance of plants and animals.

In some circumstances the geographic range of a species is tightly constrained by the existence of a suitable combination of temperature and precipitation. For example, the distributions of some forest trees conform closely to strict combinations of temperature and precipitation (Iverson and Prasad 1998). Other species occupy only a portion of the area where climate conditions are suitable for occupancy. For example, a group of birds called crossbills feed almost exclusively on the seeds of pine trees and other conifers. Even though crossbills can survive a wide range of climatic conditions, they are found only in areas that support conifer forests.

The importance of climate in understanding a species' niche can sometimes be determined by looking at how climatic conditions and the species' geographic distribution are related. If there is a broad similarity in where a species is found and where a specific combination of climatic variables is seen, then the range is said to be correlated with climate. However, the risk of relying on correlations is that important indirect relationships between species and their environments can be overlooked. For example, if climate change negatively impacts the extent of conifer forests, then crossbills may also be negatively impacted. Thus, even though crossbills may tolerate the direct effects of changes in temperature and precipitation, the effects on their food resource could cause harm.

Often the first approach used to begin studying ecological relationships is to observe the patterns of relatively undisturbed communities. These observations can reveal important relationships among species and environmental conditions, and form the basis for our understanding of the importance of climate for broad patterns of distribution.

Ecologists have sometimes collected observational data on communities for many years, producing long-term data sets showing how species and ecosystems have changed over time. Recently, researchers have used these long-term data to look for effects of the recent changes in climate. Data going back 20, 40, or even 100 years are compared with observations of climate to see how ecosystems have responded to past changes in climate. In some cases, the ecosystem itself provides a record of how it has changed over time. For example, the widths of tree rings provide information about growing conditions in the past, and fossil remains of plants and animals allow us to infer how distributions and abundances have changed. In this way, scientists can study prehistoric ecosystems dating back thousands of years or more, a discipline known as **paleoecology**. The results of analysis of long-term data sets and studies of paleoecology help ecologists understand how species responded to past periods of rapid climate change and give insights into how these same ecosystems might change in the future.

However, the initial observational approach used by ecologists is not sufficient for predicting what the results of future changes in climate will be because of the complex system of indirect effects and feedback loops characteristic of ecosystems. One way for ecologists to overcome the difficulties in evaluating systems of complex relationships is for the researcher to carefully modify one part of the ecosystem while keeping other variables constant. One can then observe the system to see how the other components change. This experimental approach provides a powerful tool for scientists to understand ecological patterns.

Ecologists studying global change have gained important insights through experimental manipulations of temperature, moisture, and CO_2. There are limits to how much we can learn about global change through experiments. Insights from experiments are reliable only if the experimental design carefully controls all of the factors that might impact the experimental subjects. Many of the interactions among species and their environment occur on large spatial scales or over long time periods. Even the largest experiments are too small to study interactions of mobile species such as birds and large flying insects. Given the limits of time, money, and other resources, some experiments cannot be done.

All of these approaches provide an enormous amount of information about how species and ecosystems respond to current variation in climate variables. While we may not always be able to conduct experiments to observe how species react, we can try to synthesize what we do know to make predictions about how ecosystems might change under future climate conditions. For one or two species and one or two environmental variables, it is possible to construct careful verbal predictions about what changes might be expected under climate change. For example, if we expect the climate in an area that currently contains large stands of coniferous forests to become so dry that the trees die of water stress, we can predict that crossbills will also disappear.

In most ecosystems, the number of relationships among species and environmental variables rapidly becomes too complex to rely on simple verbal models to predict the likely outcomes. In those cases, relationships among species and their environment can be described by mathematical equations. These sets of equations, or **ecological models**, are typically evaluated using computers. The development of the mathematical tools to evaluate ecosystems and the rapid rise in computing power has been an important development in ecology. While they require some mathematical training to understand, the advantage of models based on mathematics over verbal models is that it is possible to ensure that the logical relationships among species and their environment remains consistent.

Models have been applied to a variety of questions in global change ecology. Typically, output from models of how climate is expected to change is entered into models of how ecosystems respond to climate, producing predictions of future ecological changes.

There are two important limitations to models. First, the quality of models depends on the quality of information used to construct the model. In many cases, lack of available information about the ecological relationships among plants and animals makes it difficult to determine which variables need to be included in a model. Once the structure of a model is complete, accurate quantitative predictions depend on obtaining precise quantitative estimates of the values of ecological and climate parameters—values that are often difficult to obtain. Second, models attempt to simulate ecologists' best understanding of how variables interact, but the only way to evaluate whether models are valid is to compare what the model predicts with what really happens. Although some information about the reliability of models can be obtained by trying to re-create past events, the ultimate verification of predictions of future events depends on waiting for those events occur.

Current Research in the Field

It can take years of intensive research to understand how just a handful of environmental factors influence a single species. The following description of ongoing current research will provide the basis on which our decisions about the likely impacts of climate change will be made.

PALEOECOLOGY

Climate changes over geological time have shifted the location of suitable habitats for species, leading to shifts in species' spatial distributions. How species responded to climate changes in the past can be estimated by mapping the distribution of paleoecological data such as pollen left behind in bogs and lake sediments. Understanding how the natural vegetation has changed in response to prehistoric large climatic changes may provide information about potential future changes.

Vegetation and climate data for the last 125,000 years for the forest/steppe border of the eastern Cascade Range of North America shows variations in vegetation that are consistent with global variations in the climate system. This is evidence of climate as the primary agent of change over periods of time as long as millennia (Whitlock and Bartlein 1997).

Past changes may serve as a gauge for judging the magnitude and rate of predicted future change. Based on paleoecological data, migration rates for most species have been estimated to be between 10 and 30 km per century. This rate of migration is slower than the anticipated rates of shift in climate zones of 100 to 600 km over the next century (Solomon and Shugart 1993).

Fossil records of prehistoric animals are also a source of information on how species respond to climatic changes. One study of the remains of squid and fish at sites occupied by penguins in the Antarctic peninsula around the time of the Little Ice Age suggest that Adélie penguins may have changed their diet in response to warming and cooling cycles (Emslie et al. 1998).

Paleoecology also demonstrates the risk of species going extinct during periods of rapid climate change. During the rapid warming events 13,000 and 10,000 years ago in western Europe, records indicate that many cold-climate species of plants, insects, and large mammals were lost. Those species with wide ecological and geographical ranges were best able to survive the change, while those species with short generation times were eliminated quickly (Adams and Woodward 1992).

Paleoecological studies are limited in their ability to inform us about future changes for a number of reasons. Today, climatic changes are occurring in landscapes dominated by humans. Humans are exerting additional stress on species through habitat destruction and degradation. Migration rates calculated for prehistoric times may not reflect the future potential for species that must migrate across human-dominated landscapes. In addition, rates of climate change in coming decades are likely to be faster than those experienced in prehistoric times. The limiting factors on species migrations are difficult to infer from paleoecology and may have been limited by slower rates of climate change or by the dispersal capabilities of species.

CORRELATIONS BETWEEN CLIMATE AND ECOLOGY

Climate varies over the surface of the earth, but climate in any given spot also changes over time. In the last 100 years alone, the earth has warmed an average of 0.3° to 0.6°C (0.5° to 1.1°F). These changes are thought to affect individuals and species, ecological communities, and even how ecosystems function.

Characteristics of Individuals: Morphology and Physiology. Morphology is the form and structure of an organism, whereas physiology is the study of how those structures function. The chemical reactions that are the basis for physiology are directly influenced by temperature. At higher temperatures photosynthesis in plants can be faster and insect larvae grow more quickly. There are limits to how much these physiological processes can be speeded up. Higher rates of photosynthesis depend on adequate water and nutrients and are balanced by faster consumption of energy through respiration. At some point, temperatures become too high and these reactions can fail.

Permanent changes in individuals may result from changes in temperature. For example, in some species of reptiles, the sex of the developing embryos is determined by temperature during incubation. As temperatures at nesting areas increase more females than males are produced, changing the sex ratio (Janzen 1996).

Although warm-blooded animals are buffered from some of the effects of temperature, the morphology of mammals and birds can also change under climate change. Egg size in European pied flycatchers has increased in recent, warmer years (Järvinen 1994). Likewise, a thirty-two-year study of red deer in Norway found that males, not females. were becoming larger as climate warmed (Post et al. 1999).

Characteristics of Individuals: Phenology. Phenology is the study of periodic biological phenomena and their relationship to weather and climate. Examples include climate's relationship to the arrival of migrating birds, the start of breeding seasons, and the time when plants produce flowers. Detecting phenological changes may signal longer-term climatic changes. In addition, these studies reveal potential disruptions of interactions or relationships among species that have existed over long periods of time, such as pollination and seed dispersal.

The onset of breeding by birds in the spring is correlated with climate. Several recent long-term studies have found that birds in both North America (e.g., Dunn and Winkler 1999) and the United Kingdom (e.g., Crick et al. 1997) are breeding three to ten days earlier in the spring. The same pattern is seen in other animals such as frogs and salamanders (Bee-bee 1995).

The timing of annual migrations of birds and other animals has also been shown to be associated with climate. In New York state, thirty-nine of seventy-six bird species studied from 1903 to 1993 arrived significantly earlier in the spring in response to warmer temperatures (Oglesby and Smith 1995).

These results show that species are already responding to changes in climate. The broader implications of these studies are less well known. When species in a community respond to climate change at similar rates there may be little disruption of ecosystem structure and functioning. However, it is likely that some communities will suffer negative consequences as ecological relationships are disrupted.

For example, there are a number of long-term studies of the ecology of the great tit, a relative of the North American chickadees, in the woodlands of Europe. The birds depend on an ample supply of caterpillars to successfully raise their chicks. The time when caterpillars become abundant in the spring depends on when the oak trees in the forest produce their leaves. In the Netherlands, warmer springs over the last twenty-three years have led to earlier production of leaves, which in turn has resulted in the peak abundance of caterpillars coming nine days earlier. However, breeding of the local great tits has not started earlier so that many parents are trying to raise their offspring when food availability has started to decline (Visser et al. 1998).

Characteristics of Populations: Range. Increases in temperature and changes in precipitation are likely to lead to changes in the ranges and abundances of some species as current habitats become unsuitable and new habitats arise.

In North America and Europe, the ranges of butterfly species have shifted north, coinciding with recent increases in temperature. Camille Parmesan (1996) focused on the distribution of Edith's checkerspot butterfly in western North America. Because long-term data on the distribution of this species were not available, Parmesan used carefully collected information from museums, butterfly collectors, and publications to determine the historical locations of Edith's Checkerspot populations. She then revisited these historical locations and found that many populations in the south of the species range had gone extinct, leading to a northward retraction in the species' range (Parmesan 1996). Parmesan and a group of collaborators used a similar approach to document northward shifts in European butterflies (Parmesan et al. 1999).

There are close links between climate and geographic ranges for many species of birds and recent climate changes are associated with range changes (Thomas and Lennon 1999; Venier et al. 1999). In the Antarctic, gentoo and chinstrap penguins have expanded their ranges southward over the last fifty years in coincidence with regional warming shifts (Emslie et al. 1998).

Characteristics of Populations: Abundance. The abundance and the persistence of some species have been linked to climate change. Amphibians such as frogs and salamanders are susceptible to climate change because they are sensitive to moisture changes and often use more than one habitat type and food type in their lifetimes. Between 1979 and 1989 in Puerto Rico, populations of frogs declined, corresponding with low precipitation events over the same period (Stewart 1995).

In some cases population changes will be severe enough to result in the local extinction of a species. For decades, the annual breeding of the golden toad in the Monteverde cloud forest reserve of Costa Rica presented the spectacle of the sudden appearance of thousands of bright orange toads on the forest floor. The sudden and unexpected extinction of this species in the late 1980s was associated with unusually warm and dry conditions that also resulted in the local extinction of the harlequin frog and declines in populations of other frogs (Pounds et al. 1999).

The fact that species will not respond in the same way to climate change will result in changes in ecosystem structure. As water temperatures have increased, populations of invertebrates along the rocky intertidal zone of the Pacific Ocean in California have changed in recent years, with warm-water species becoming more abundant and cold-water species declining (Barry et al. 1995). As a result, the ecological community is different from when the same locations were first studied in the 1930s.

Characteristics of Communities and Ecosystems. Effects of changing climate on individuals and species have implications for higher levels of ecological organization. The effects of climate change have been linked to changes in aquatic and terrestrial communities.

Climate change can affect both the surface temperature of ocean waters and the patterns of currents that transport nutrients. Worldwide, coral reefs are threatened by an increase in the frequency of coral bleaching, a syndrome that occurs when coral expel the algae that live inside them. These bleaching events become more frequent when ocean temperatures are higher than normal (Huppert and Stone 1998). Open ocean ecosystems are also affected by climate. For example, there has been an 80% decrease in the biomass of macrozooplankton off the coast of California, coinciding with a long-term increase of 0.8° to 1.5°C (1.4° to 2.7°F) in ocean temperature since 1951 (Roemmich and McGowan 1995). The change in water temperature may have limited the inputs of nutrients into the ecosystem, contributing to a decrease in the macrozooplankton. Such effects may flow up the food chain and may be responsible for population declines in Pacific Ocean seabirds (Veit et al. 1997).

Temperature and precipitation are strong determinants of vegetation patterns across the globe. In arid regions of the southwestern United States, changes in precipitation have shifted some sites from grassland to desert shrubland (Brown et al. 1997). In northeast Colorado, the primary ground cover is shortgrass prairie, which has declined in annual net primary productivity and abundance with a rise in average temperatures of 1.3°C (2.3°F) since 1970 (Alward et al. 1999). On the other hand, the net primary productivity of broadleaf plants has increased with warming (Alward et al. 1999).

Productivity of ecosystems throughout the Northern Hemisphere has increased since the 1960s. Measurements of changes in growing season, such as seasonal changes in the CO_2 cycle, remote sensing data, and observations of plant phenology, reveal a lengthening in the growing season of between ten and twelve days, thought to be responsible for the higher productivity (Myneni et al. 1997; Menzel and Fabian 1999).

EXPERIMENTAL STUDIES OF CLIMATE CHANGE

Experiments are one of the most powerful tools available to ecologists. Carefully designed and executed experimental manipulations allow scientists to overcome many of the limitations of correlational studies, and to document cause-effect relationships with a high degree of certainty.

The logistic difficulties involved in manipulating temperature and other climate variables over even a moderate area means that experiments have been most widely used to study questions about the physiology and behavior of individuals. By enclosing plants and small animals in climate chambers, temperature, moisture, and CO_2 concentrations can be carefully controlled.

Results of these studies have shown that the response of plants to global change will depend on the combined effects of changes in temperature, precipitation, and CO_2 concentrations. CO_2 is a key resource for photosynthesis in plants, and rising concentrations in the atmosphere can increase plant growth. However, plants can respond to increases in CO_2 concentrations only if they have abundant water and nutrients. Higher plant production can be fueled only by increased precipitation, but in many areas rainfall is expected to decline. Even in areas where precipitation remains constant, higher temperatures will increase water use and lead to stress (IPCC 1996b).

Plants important in agriculture and forestry will be affected by changes in climate, and these changes will have a direct impact on humans (IPCC 1996b). Regions with abundant precipitation may have higher yields as a result of higher levels of CO_2, although this may require higher inputs of expensive fertilizers to support rapid growth. In other areas higher temperatures and water stress may force farmers and foresters to switch to more tolerant species. In natural ecosystems, species that are best able to adjust to climate change will have a competitive advantage over others.

One of the biggest limits to the experimental approach for understanding climate change is the logistical difficulty involved in enclosing organisms in controlled climate chambers. Chambers holding single individuals of small species are relatively easy to build and even room-size chambers are increasingly common. Although these approaches are invaluable for studying the physiology of individuals, it becomes more difficult to understand how individuals will interact with other components of ecosystems.

Some experimental research is being conducted on the ecosystem-level effects of elevated CO_2 and its interactions with other environmental factors. The impact of higher CO_2 levels on vegetation is being studied in a variety of ecosystems. These studies, referred to as free-air carbon dioxide enrichment (FACE) experiments, typically surround a moderate-sized outdoor area (approximately 1 to 30 m in diameter) with a system of pipes designed to deliver a constant stream of CO_2 to the surrounding vegetation (McLeod and Long 1999). In this way, plants are allowed to interact with each other and with other components of their environment.

Many FACE experiments have resulted in increases in plant productivity. For example, one study exposing a cotton crop to FACE (550 ppm of CO_2) on a large-scale production environment resulted in increases in biomass and harvestable yield of 37% to 48%. These results are thought to be attributable to increased early-leaf area, more profuse flowering, and a longer period of fruit retention (Mauney et al. 1994).

Not all species respond equally to changes in CO_2 concentrations. Species-specific responses to CO_2 produce shifts in ecosystem structure, even in the absence of changes in precipitation and temperature as plant communities shift to favor species that use CO_2 efficiently. Experimental studies of the effects of changes in temperature and precipitation on the insects that feed on plants emphasize the additional ecological complexities that have yet to be fully understood. For example, when temperature is experimentally increased by 1°C (1.8°F) spittlebug densities increase by 157% (Whittaker and Tribe 1998), resulting in added stress on plants.

Trade-offs exist in the design of ecological experiments. Carefully controlled laboratory experiments are difficult to extrapolate to field conditions because they are typically done under ideal conditions and cannot include all of the environmental interactions found in ecosystems. Results from lab experiments may not be applicable to plants that may be grown under suboptimal conditions of low precipitation or nutrients. As experiments incorporate more variables, it becomes more difficult to control all of the nonexperimental variables, making results difficult to interpret. For example, field studies may incorporate more realistic environmental conditions, but the enclosures for field-grown plants tend to modify the plant's environment in unexpected ways. Thus, experiments need to be combined with careful observations of unmanipulated systems.

PREDICTIONS USING MODELING TECHNIQUES

Paleoecology, observational studies, and experiments all provide important information about how species and ecosystems respond to climate variables. The ultimate goal of ecologists interested in minimizing the impact of future climate change is to be able to predict how ecosystems will change in the coming decades.

How individual species respond to climate change is of interest both because many species are valued by humans (e.g., valuable forest trees, endangered and threatened wildlife) and because characteristics of species influence the structure and functioning of the ecosystems. Predictions of the effects of climate change on individual species are based on the relationship between a species' current distribution and a set of climate parameters. This method has been used to predict shifts in the distribution of fish, aquatic insects, birds, and vegetation.

Fish cannot regulate their internal temperatures so they seek habitat close to the optimum temperature for growth, foraging success, and protection. Fish species have different preferred temperatures and thus their ranges may expand or shrink as a result of temperature changes. If air temperatures increase 4°C (7.2°F), smallmouth bass and yellow perch in North America are expected to be able to move as much as 500 km farther north (Shuter and Post 1990). Overall, habitat in the United States for cold freshwater species such as trout is predicted to decline by 47% due to warmer temperatures (Eaton and Scheller 1996).

Climate modeling has also been used to predict the potential effects of climate change on habitat availability for threatened and endangered species. One example is the study by Dexter et al. (1995) in which the potential effects of climate change on Australia's threatened vertebrate populations were examined. Using three scenarios of climate change, predictions were made of shifts in species' ranges. Loss of 90% or more of core habitat was projected for between three and thirteen species by 2030. Another forty-six to fifty-five threatened species were projected to lose some potential core habitat, while two to eleven of the species were projected to have the same amount or more climatically suitable habitat available to them by 2030.

Modeling approaches that relate average climatic variables to distributions of **biomes** or species to forecast changes in individual species or communities often do not consider nonclimatic factors. For example, competition with other species, barriers to migration, and distributions of pests and pathogens may have significant effects on distributions of species. As models do incorporate more environmental interactions, they become more difficult to interpret. Because some species will likely respond more strongly to the climate and others to factors such as vegetation, future studies need to examine both factors in order to understand how the actual ranges might change and what the rates of change might be. Two major limitations in the development of models are uncertainty about how climate, especially rainfall, will change for specific areas and the difficulty in understanding all of the ecological interactions that govern the distribution and abundance of species.

Conclusion

The realization that the earth is undergoing a rapid change in climate has led to the increasing need for ecologists to provide information about how valued species and ecosystems will respond. While it is clear that climate is of central importance in determining the abundance and distribution of species, the complexity of natural systems means that it is still difficult to make precise predictions about changes that can be expected in any specific area. As the types of studies described here become more sophisticated, predictions will become more reliable, but the very nature of ecosystems means that there will always be a high probability of unanticipated consequences occurring.

Human changes to the landscape are already having a deleterious effect on biological diversity. Climate change is likely to confound the future of biological diversity to an even greater extent. Evidence suggests that climate changes will cause individual species to react differently. Species' distributions will change at different rates, producing new animal and plant communities. Changes in ranges of many species will result, along with disruptions of natural communities and extinction of species incapable of adaptation. Evidence from the fossil record indicates climate change has been an important cause of extinctions in the past.

The loss of species, especially those that are already threatened or endangered, will likely accelerate because of reductions of suitable habitat and changes in climate that are faster than species' abilities to adjust. Although some species may become extinct, climate changes may favor the spread of invasive weedy species that are already a leading factor causing alterations in the structure and functioning of ecosystems. Invasive species do not merely compete with or consume particular native species, they change the nature of the ecosystem in which all natives survive.

Reduction in biological diversity can have profound effects on the capacity of an ecosystem to provide ecosystem services (Naeem et al. 1999). Ecosystems provide services to society including clean water, pure air, soil formation and protection, pest control, foods, fuel, fibers, and drugs (Daily 1997). The loss of biodiversity reduces the flexibility of species to adjust to environmental changes such as climate changes and narrows the options available for the rehabilitation of specific habitats.

The rapid development in our understanding of the importance of climate change for ecosystems emphasizes the importance of developing

Ecosystems in Danger

strategies to cope with future climate change. Natural variation in climate and the difficulties in minimizing human influence on climate systems mean that some degree of climate change is inevitable. In addition to the current need for ecologists to provide information about how ecosystems are likely to change, there is an increasing need for ecologists to develop strategies to minimize negative impacts on important ecological services.

REFERENCES

Adams, J. M., and F. I. Woodward. 1992. The past as a key to the future: The use of paleoenvironmental understanding to predict the effects of man on the biosphere. *Advances in Ecological Research* 22:257–314.

Alward, R.D., J. K. Ketling, and D. G. Milchunas. 1999. Grassland vegetation changes and nocturnal global warming. *Science* 283:229–31.

Barry, J. P., C. H. Baxter, R. D. Sagarin, and S. E. Gilman. 1995. Climate related, long-term faunal changes in a California rocky intertidal community. *Science* 267:672–75.

Beebee, T. J. C. 1995. Amphibian breeding and climate. *Nature* 374:219–20.

Brown, J. L., T. J. Valone, and C. G. Curtin. 1997. Reorganization of an arid ecosystem in response to recent climate change. Proceedings of the National Academy of Sciences, USA 94:9729–33.

Crick, H. Q. P., C. Dudley, D. E. Glue, and D. L. Thomson. 1997. UK birds are laying eggs earlier. *Nature* 388:526.

Daily, G. C., ed. 1997. *Nature's services: Societal dependence on natural ecosystems.* Washington, D.C.: Island Press.

Dexter, E. M., A. D. Chapman, and J. R. Busby. 1995. *The impact of global warming on the distribution of threatened vertebrates.* Australian Environment Online, http://www.erin.gov.au/life/end_vuln/animals/climate/climate_change/cc_con.html.

Dunn, P. O., and D. W. Winkler. 1999. Climate change has affected breeding date of tree swallows throughout North America. Proceedings of the Royal Society of London 266:2487–90.

Eaton, J. G., and R. M. Scheller. 1996. Effects of climate warming on fish thermal habitat in streams of the United States. *Limnology and Oceanography* 41:1109–15.

Emslie, S. D., W. Fraser, R. C. Smith, and W. Walker. 1998. Abandoned penguin colonies and environmental change in the Palmer Station area, Anvers Island, Antarctic Peninsula. *Antarctic Science* 10:257–68.

Huppert, A., and L. Stone. 1998. Chaos in the Pacific's coral reef bleaching cycle. *American Naturalist* 152:447–59.

Intergovernmental Panel on Climate Change (IPCC). 1996a. *Climate change 1995: The science of climate change,* edited by J. T. Houghton, L. G. Meira Filho, B. A. Callander, N. Harris, A. Kattenberg, and K. Maskell. New York: Cambridge University Press.

———. 1996b. *Climate change 1995: Impacts, adaptations and mitigation of climate change: Scientific-technical analyses,* edited by R. T. Watson, M. C. Zinyowera, and R. H. Moss. New York: Cambridge University Press.

Iverson, L. R., and A. M. Prasad. 1998. Predicting abundance of 80 tree species following climate change in the eastern United States. *Ecological Monographs* 68:465–85.

Janzen, F. J. 1994. Climate change and temperature-dependent sex determination in reptiles. Proceedings of the National Academy of Sciences, USA 91:7487–90.

Järvinen, A. 1996. Correlation between egg size and clutch size in the pied flycatcher *Ficedula hypoleuca* in cold and warm summers. *Ibis* 138:620–23.

Mauney, J. R., B. A. Kimball, P. J. Pinter, R. L. Lamorte, K. F. Lewin, J. Nagy, and G. R. Hendrey. 1994. Growth and yield of cotton in response to a free-air carbon dioxide enrichment (FACE) environment. *Agricultural and Forest Meteorology* 70(1–4):49–67.

McLeod, A. R., and S. P. Long. 1999. Free-air carbon dioxide enrichment (FACE) in global change research: A review. *Advances in Ecological Research* 28:1–56.

Menzel, A., and P. Fabian. 1999. Growing season extended in Europe. *Nature* 397:659.

Myneni, R. B., C. D. Keeling, C. J. Tucker, G. Asrar, and R. R. Nemani. 1997. Increased plant growth in the northern high latitudes from 1981–1991. *Nature* 386:698–702.

Naeem, S., F. S. Chapin III, R. Costanza, P. R. Ehrlich, F. B. Golley, D. U. Hooper, J. H. Lawton, R. V. O'Neill, H. A. Mooney, E. S. Osvaldo, A. J. Symstad, and D. Tilman. 1999. Biodiversity and ecosystem functioning: Maintaining natural life support processes. *Issues in Ecology No. 4.* Washington, D.C.: Ecological Society of America.

Oglesby, R. T., and C. R. Smith. 1995. Climate changes in the Northeast. In *Our living resources: a report to the nation on the distribution, abundance, and health of U.S. plants, animals, and ecosystems,* edited by E. T. LaRoe, G. S. Farris, C. E. Puckett, P. D. Doran, and M. J. Mac, 390–91. Washington, D.C.: U.S. Department of the Interior, National Biological Service.

Parmesan, C. 1996. Climate and species' range. *Nature* 382:765–66.

Parmesan, C., N. Ryrholm, C. Stefanescu, J. K. Hill, C. D. Thomas, H. Descimon, B. Huntley, L. Kaila, J. Kullberg, T. Tammaru, W. J. Tennent, J. A. Thomas, and M. Warren. 1999. Poleward shifts in geographical ranges of butterfly species associated with regional warming. *Nature* 399:579–83.

Post, E., R. Langvatn, M. C. Forchhammer, and N. C. Stenseth. 1999. Environmental variation shapes sexual dimorphism in red deer. Proceedings of the National Academy of Sciences, USA 96:4467–71.

Pounds, J. A., M. P. L Fogden, and J. H. Campbell. 1999. Biological responses to climate change on a tropical mountain. *Nature* 398:611–15.

Roemmich, D., and J. McGowan. 1995. Climatic warming and the decline of zooplankton in the California Current. *Science* 267:1324–26.

Shuter, B. J., and J. R. Post. 1990. Climate, population viability and the zoogeography of temperate fishes. *Transactions of American Fisheries Society* 119:316–36.

Solomon, A. M., and H. H. Shugart, eds. 1993. *Vegetation dynamics and global change.* London: Chapman and Hall.

Stewart, M. M. 1995. Climate driven population fluctuations in rain forest frogs. *Journal of Herpetology* 29:437–46.

Thomas, C. D., and J. J. Lennon. 1999. Birds extend their ranges northwards. *Nature* 399:213.

Veit, R. R., J. A. McGowan, D. G. Ainley, T. R. Wahls, and P. Pyle. 1997. Apex marine predator declines ninety percent in association with changing oceanic climate. *Global Change Biology* 3:23–28.

Venier, L. A., D. W. McKenney, Y. Wang, and J. McKee. 1999. Models of large-scale breeding-bird distribution as a function of macro-climate in Ontario, Canada. *Journal of Biogeography* 26:315–28.

Visser, M. E., A. J. van Noordwijk, J. M. Tinbergen, C. M. Lessells. 1998. Warmer springs lead to mistimed reproduction in great tits (*Parus major*). Proceedings of the Royal Society of London. B 265:1867–70.

Whitlock, C., and P. Bartlein. 1997. Vegetation and climate change in northwest America during the past 125 kyr. *Nature* 388:57–61.

Whittaker, J. B., and N. P. Tribe. 1998. Predicting numbers of an insect (*Neophilaenus lineatus: Homoptera*) in a changing climate. *Journal of Animal Ecology* 67:987–91.

SUGGESTED READINGS

Daily, G. C., S. Alexander, P. R. Ehrlich, L. Goulder, J. Lubchenco, P. A. Matson, H. A. Mooney, S. Postel, S. H. Schneider, D. Tilman, and G. M. Woodwell. 1997. Ecosystem services: Benefits supplied to human societies by natural ecosystems. *Issues in Ecology No. 2.* Washington, D.C.: Ecological Society of America. Also available online: http://esa.sdsc.edu/issues.htm.

Intergovernmental Panel on Climate Change (IPCC). 1996. *Climate change 1995: Impacts, adaptations and mitigation of climate change: Scientific-technical analyses,* edited by R. T. Watson, M. C. Zinyowera, and R. H. Moss. New York: Cambridge University Press.

Kareiva, P. M., J. G. Kingsolver, and R. B. Huey. 1993. Biotic interactions and global change. Sunderland, Mass.: Sinauer Associates.

Mac, M. J., P. A. Opler, S. T. A. Pickett, C. E. Haecker, P. D. Doran. 1998. Status and trends of the nation's biological resources. Vol. 1. Reston, Va.: U.S. Department of the Interior, U.S. Geological Survey. Also available online: http://biology.usgs.gov/s+t/SNT/index.htm.

Naeem, S., F. S. Chapin III, R. Costanza, P. R. Ehrlich, F. B. Golley, D. U. Hooper, J. H. Lawton, R. V. O'Neill, H. A. Mooney, E. S. Osvaldo, A. J. Symstad, and D. Tilman. 1999. Biodiversity and ecosystem functioning: Maintaining natural life support processes. *Issues in Ecology No. 4.* Washington, D.C.: Ecological Society of America. Also available online: http://esa.sdsc.edu/issues.htm.

The Pacific Institute for Studies in Development, Environment, and Security provides a comprehensive bibliography of climate change impact on biodiversity. Available online: http://www.pacinst.org/wildlife.html.

Peters, R. L., and T. E. Lovejoy. 1992. *Global warming and biological diversity.* New Haven, Conn.: Yale University Press.

U.S. Environmental Protection Agency. *The EPA global warming site,* http://www.epa.gov/globalwarming/index.html.

CHAPTER

4

Cultural Paradigms and Challenges

An
THROPOLOGICAL
ERSPECTIVE ON
LIMATE CHANGE

Celeste Ray

In the warm summer of 1991, a glacier near the Alpine pass of Tisenjoch began melting at an unprecedented rate. Just over the Italian border from Austria, hikers discovered the body of a 5,000-year-old man, frozen since Europe's Copper Age. Naturally mummified, the "Ice Man" retained his skin, internal organs, and even traces of his last meal. In 1999, another ice man surfaced from a glacier in the Canadian Yukon. Hunters and hikers frequently passed his find site in the Tatshenshini-Alesk Wilderness Park, but only with glacial melting did the almost 600-year-old body and his remarkably intact clothing and tools reemerge. The same year also witnessed the recovery of a late-Pleistocene woolly mammoth that had been perfectly preserved in the permafrost of northern Siberia for 20,000 years.

The recovery of such ancient remains at such geographically dispersed sites suggests global warming is already at work in dramatic ways. By revealing new finds, global warming might seem beneficial for archaeologists and paleontologists, but their main interest in climate change is in understanding how past societies culturally responded to similar situations and what we can learn from their successes and failures.

In America, **archaeology** is a subdiscipline of **anthropology**, which studies all humankind and cultures across space and time. We can define **culture** as all learned behaviors and beliefs: what we think and do that is not instinctual. Our culture shapes, but does not determine, our worldview. Individuals, of course, have unique ways of thinking and acting, but those within a society or community are **enculturated** to share common cultural models (ways of perceiving and acting in the world). **Ecological anthropology** draws on all subfields of our discipline to consider past and present relationships between societies and their environments as mediated through culture. **Ecology** examines relationships between environments and the organisms that dwell within them. When anthropologists pursue ecological studies they examine the environment as people are affected by it, as they use it, as they seek to understand it, and as they modify it. Ecological anthropologists are, then, particularly interested in the ways global climate change will alter human cultures, settlement, interaction, and use of natural resources. Ecological anthropologists are also concerned with how local communities will interpret and respond to environmental problems on a global scale. Accepting that the present rate of global warming is human induced, anthropologists suggest that we can develop workable policies for its reduction only if we understand why people value activities that damage the environment and what would make them alter their views.

Paradigms in the Field

Through many and varied approaches, ecological anthropology studies the interplay between human population behavior and environmental variables through time (Ellen 1982). **Cultural ecology** considers how the culture of a human group is adapted to and shaped by the natural resources of an environment and to the presence of other human groups. Cultural ecologists concerned with climatic change consider how the interrelationships between subsistence strategies (ways of acquiring food and resources from the environment), settlement patterns, and social life can and do evolve to meet environmental challenges. **Behavioral and evolutionary ecology** considers how humans evolve behaviorally or physically to a given environment and in relation to changes in that environment over time. **Ethnoecology** describes people's own conceptual models of their environment: how they classify and interact with plants, animals, landforms, and water bodies, and how different cultures attribute significance and divine attributes to environments. ("Ethno" means people or folk, so that ethnoecology examines folk models or cultural models of environments.) Ethnoecological accounts can give us insights from other cultures for tackling environmental problems. Unlike most scientific disciplines, ecological anthropology examines how feelings, traditions, and religious values affect how humans interact with their environments, how environments are regarded and treated, and how their health may be maintained. Each of these approaches within ecological anthropology can help us see how people perceive, understand, and respond to climate change, and therefore help us to create effective strategies for changing behaviors, subsistence techniques, and natural resource consumption practices that contribute to global warming.

Examining human actions that contribute to environmental crises, we must remember that what people say they do and what they actually do are not always congruous. On surveys or questionnaires, respondents may answer in ways they think are appropriate rather than in ways that reveal their actual lifestyles. Anthropologists rely on interviews and supplement verbal accounts by observing the day-to-day activities of communities and individuals in the workplace, in the home, and in public arenas through a research strategy called **ethnography**. For example, in a study observing consumption patterns and comparing finds with ethnographic accounts of what people thought they consumed, people consistently underreported their actual consumption (Harrison et al. 1992). By actually observing daily practices of individuals and communities in the consumption of **fossil fuels** (the main contributor to **global warming**), we can distinguish between necessity use and lifestyle use and be better able to design energy efficiency measures and lifestyle changes that the public will find acceptable and follow.

When we study how humans have interacted with their environments in the past, anthropologists turn to **ethnohistory**. To examine past ecosystems and human responses to climatic or environmental change, ethnohistorians study primary written sources (e.g., government documents, diaries, letters), oral histories (e.g., stories about droughts or pest invasions) and even artwork and photography that reveal changes in a landscape over time (Crumley 1994). The type of anthropological ecology that expressly studies past human cultures' responses to, and impacts on, their natural environments is **historical ecology**. Drawing on archaeology, ethnohistory, and ethnography, historical ecology employs concepts from both the natural and social sciences to explore the complex chains of mutual causation in human-environmental relations over time (Crumley 1994). Historical ecology is especially well positioned to examine how people have responded, or failed to respond, to climatic changes in the past and what their choices have meant for the fate of their societies. Historical ecology gives us historical analogues by which to develop our own strategies for coping with current and future environmental crises.

Of the many perspectives within ecological anthropology, this chapter will draw strongly from ethnoecology and historical ecology in considering the varied and dynamic ways through which past and present societies have conceived of their environments, human stewardship of natural resources, and environmental and climatic change. While contrasting **biocentric** and **anthropocentric** visions of the human-environmental relationships, this discussion challenges the myth of primitive ecological wisdom and asks what such myths reveal about the cultural models employed by both environmentalists and those who still imagine climatic change to be a "doom and gloom" fiction. Employing historical case studies and contemporary

Cultural Paradigms and Challenges   85

examples, the chapter explores how cultural understandings and evaluations of environmental crises have contributed to human attempts to address climate change.

Discussion

Cultural models shape our response to environmental crises. How then do we culturally conceive of the environment? Any given society has many varied perceptions of the environment and an even greater diversity of perspectives exists *between* cultural traditions. Such variances will inevitably influence how different cultures around the world will adapt to the climatic changes associated with global warming. Even scientists trained within the same broad tradition of Western science conceive of "nature" in different ways. Anthropologist Kay Milton (1996) notes that ecologists who perceive nature as robust or even capricious are less likely to perceive the risks involved in our use of the environment or our responsibilities toward it. In other words, if we see the environment as invincible, we are less likely to see ourselves as its stewards or protectors.

Cultural models of the environment are not static; they change over time with environmental changes, with population dynamics, with religious beliefs, and with new technologies. For millennia, if not millions of years, our chief practical worry as humans has been protecting ourselves from other animals, the elements, and seasonal weather changes. We have spent 99% of human history hunting and gathering what we needed to exist on a daily basis. We learned to domesticate plants and animals only with the beginning of the warmer **Holocene** epoch about 10,000 years ago, allowing population numbers to soar and villages and cities to develop around a dependable food supply. We have harnessed the forces of nature for our own ends in the industrial and technological ages only within the last three centuries. We have experienced the brief euphoria of controlling nature and now find that our inventive control may have repercussions we did not predict and cannot govern.

EVALUATING NATURE: ANTHROPOCENTRISM AND BIOCENTRISM

Archaeologist Brian Fagan (1999) reminds us that "global warming is nothing new for humanity." Our remote forebears experienced dramatic climatic shifts from ice ages to warm periods; however, when our hunter-and-gatherer ancestors experienced "warm-ups" the world's population "numbered in the tens of thousands rather than the billions." People could move into other territories as the animals they hunted and the plants they foraged expanded or retreated in response to climatic conditions. Once people began settling in agricultural communities, they became tied to location, landownership, and tradition and could no longer be so flexible in coping with climatic changes. Therefore, warm "climatic optimums" and "mini-ice ages" would seem to have much more powerful impacts on settled, rather than nomadic, societies.

Most state-level societies of the last millennia, and even the world's first state-level societies in Mesopotamia, have viewed nature as a thing to be subdued and conquered (Ferkiss 1993). Nature presents a challenge or a hindrance to human endeavors and economic development. This attitude explains why we are so reluctant to believe that human actions can impact our environments in irreversible ways. It also explains our denial of how rapidly our mineral extraction, energy production, and toxic chemical waste "disposal" are negatively changing our environment. A. M. Mannion (1991) notes that "Both acidification, especially significant in the Northern Hemisphere where most industrialisation [sic] has taken place, and the enhanced greenhouse effect are products of the post-1700 industrial era." Despite such recent and dramatic changes in environmental conditions, predominant cultural models in industrialized societies continue to evaluate the environment as all-powerful and self-renewing.

Our perceptions of human-environmental relationships derive in part from the classification schemes of the eighteenth-century botanist Carolus Linnaeus who, in accordance with **Judeo-Christian** interpretations, placed humans at the top of a hierarchy of plants and animals. Nineteenth-century Darwinian concepts of "survival of the fittest" supported

Cultural Paradigms and Challenges

this anthropocentric vision of the world. Perhaps we were animals, but we alone had organized ourselves into civilizations and subdued the earth. Therefore we must be the "fittest" and could do as we pleased to other nonhuman species and the land. It is reductionistic to suggest these ideas singularly shaped our perceptions of nature. We also have the cultural idea of "Mother Nature," and more recently the rebirth of the ancient Greek "Gaia" as a theory that alternately suggests our "mother" takes care of us. However, it is the ideas of the **Enlightenment** and the nineteenth century that have guided governments, entrepreneurs, and policy makers as they pushed for "progress." Progress in the nineteenth and twentieth centuries envisioned humans as emerging from environmental dependency to control of nature. Anthropologists once ranked cultures from the savage, to the barbaric, to the civilized depending on the degree to which they had removed themselves from nature's "bonds."

Esteeming an "escape" from the so-called bonds of nature tends to deny human stewardship responsibilities as a part of nature. Our conceptions of nature tend to be local or regional, and we rightly worry about predicting and limiting the damage of short-term events such as earthquakes, tornadoes, or even thunderstorms. However, thinking about the long-term damage that human-induced climatic change might cause for future generations and the fate of societies on a global scale is still a relatively new dimension to our worldviews and one that many laypersons and scientists alike resist. One reason such resistance is so strong in the most industrialized societies is that Western (and Middle Eastern) perceptions of nature anthropocentrically set humans *apart* from nature. When this cultural perception enters the realm of policy making, we speak of the preservation or conservation of nature, not for itself and other creatures, but for our children or because we need its resources to perpetuate our lifestyle. Our language reveals this; we say the "marsh" needs draining when its presence complicates a development project, but we call the same area "wetlands" when we perceive it as valuable.

In the United States, we are interested in preserving certain landscapes because we like looking at them or hiking through them. The most common way to draw support for preservation goals has been to appeal to our aesthetic tastes. Landscape artists such as George Catlin, Albert Bierstadt, and Frederic Church used their work in nineteenth-century efforts to establish the United States' first national parks. Groups like the Sierra Club and The Nature Conservancy continue to attract support with visual imagery. However, just like our cultural worldviews, our perceptions of natural beauty change over time. In eighteenth-century Europe, the Alps and the Highlands of Scotland were considered wild, wicked, and ungodly places. By the nineteenth century, shifts in cultural perspectives encouraged by romanticism and booming tourism industries reevaluated these same places as "sublime," "picturesque," and of great "moral value." Landscape elements we perceive as beautiful may be quite different from those another culture would likewise value, partly because most all landscapes reveal culturally influenced human modifications.

We have altered our environments and they have altered us over time. Indeed, we hardly realize that many of the landscapes we admire are not "natural" but humanly manipulated to fit our own cultural frames for beauty or utility. Roy Ellen (1982) reminds us that "Few environments are uninfluenced by humans and almost all landscapes are culturally modified to some extent. . . . local ecosystems are everywhere relatively recent." Because we objectify nature, we detach ourselves from it and fail to see our impact on local, much less global scales. We also then fail to acknowledge our responsibilities for its maintenance.

In contrast with the Western tendency toward anthropocentric conceptions of an ever-giving environment, many cultures see humans in a reciprocating relationship with nature; it gives as long as we take carefully. Some cultures, and many environmentalists, hold more biocentric views in which humans are a part of nature, and as such, respect the "natural" rights of nature, not just those of humankind. Biocentric thought requires us to think beyond hierarchical models to also consider **heterarchically** how environmental elements ranked most important by humans (or at the top of the food chain) may not always be the most significant in shaping the health of an ecosystem.

THE MYTH OF PRIMITIVE ECOLOGICAL WISDOM

Cultural visions of natural resources as abundant, renewable, and even existent to be exploited by humans has led to an international market of local natural resources accompanied by environmental crises such as the loss of biodiversity, acid rain, and increasing **greenhouse gas** emissions, which have become the "tragedy of the global commons" (McCay and Acheson 1990). The responsibility for predicted climate change is usually laid at the feet of industrialized nations who anthropocentrically, and often ethnocentrically, acquire and consume the largest quantities of the world's natural resources. While deserved in many ways, such a critique often simplistically fails to acknowledge that demand is met by suppliers and that people did degrade their environments before industrialism.

For many environmentalists, turning to more biocentric perspectives of the environment means not simply evolving and expanding our ideas, but rejecting Western traditions to embrace other cultures' perceptions of nature. Many people uncritically accept a myth of primitive ecological wisdom: that nonindustrial societies possess a "pure" ecological wisdom that enables them to live in harmony with their environments. However, many nonindustrial societies have not used their environmental resources in a sustainable way. For example, in adapting to a new environment, the Polynesians who settled New Zealand a millennium ago caused several centuries of environmental destruction including the extermination of flightless birds (Harre et al. 1999). While the cassowary bird still thrives in the territory of the Karam people of the New Guinea Highlands (because they view the bird as their metaphorical kin), the Karam's neighbors hunted the bird out of existence (Milton 1996). Believing trees can cause human deaths if mistreated, the Dogon of Mali in West Africa respect trees and their environment in the same ways as they show deference to elders and superiors in their social hierarchy (Milton 1996). To fell an entire tree is a serious decision that requires group agreement and often an offering. Though the Dogon revere their environment, they feel it is more powerful than themselves and that, because it cares for them, they are not responsible for replenishing what they have used. What can be learned from these examples is that nonindustrial societies have not always maintained practices and traditions compatible with preservation, or even conservation strategies, and do not necessarily illustrate the responses needed to address climatic changes on national, regional, or global scales today.

Anthropologists are more interested in why people embrace myths than whether or not they are true. The myth of primitive ecological wisdom suggests that people living in "civilization" separated from nature have lost the way to becoming ecologically wise, or believe they have lost their way. It also reveals insecurities and a lingering willingness to romanticize in current cultural models of Western environmentalism. We must consider how such a myth delegates, or fails to delegate, responsibility for environmental crises and for future policies to address them. For instance, how might the use of such myth in environmental discourse impose a sense of collective guilt (however merited) on industrialized nations and thereby alienate (rather than attract) their policy makers to agreements and practices to lower greenhouse gas emissions and address human-induced climate change? True or false, the myth of primitive ecological wisdom has at least interested individual citizens and some policy makers in what smaller-scale, less materialistic, nonindustrial societies do have to teach us about environmentally friendly values and practices.

The world's wealthiest nations, most responsible for anthropogenic warming, have developed and exported standards of living that shape an increasingly global culture. Though we now realize what achieving our way of life has cost the environment, we in the technologically and materially wealthy countries cannot prevent development elsewhere. How can we assist developing nations, as far as they ask us to, without also globally exporting the practices that raise greenhouse gas emissions? To create policies people will willingly follow, we must employ cross-cultural knowledge of benign environmental practices while remaining conscientious of traditional family and community structures, labor division, and localized subsistence strategies. We must learn about a society's use of biological resources, its environmental ethos, and religious beliefs. We can use this

Cultural Paradigms and Challenges   91

combined knowledge to preserve both cultural and environmental diversity while satisfying people's short-term material desires. Though a tall order, it is a variety of perspectives and strategies that can perhaps shape the most effective stewardship of our common home.

RELIGIOUS WORLDVIEWS AND CULTURAL RESPONSES TO ENVIRONMENTAL CRISES

Anthropologists are concerned with understanding how people might respond to the variety of problems currently theorized to accompany global warming: increased El Niños, crop failures, famine, epidemics, flood, and drought. Individual occurrences of these types of problems lead people to look to their religion and/or their governments to "fix" the uncontrollable. Human responses to environmental crises and epidemics in the past have often involved extreme conservatism and irrational convolutions of religious beliefs that often yielded even more detrimental practices. Policy makers and strategists who attempt to limit societal disruption and hardship in current environmental crises and who try to educate the public about lifestyle changes related to the possible impacts of global warming can benefit by considering what type of reactions and behaviors they might expect cross-culturally.

Brought to humans by flea-carrying rats, the bubonic plague took its heaviest toll in Europe between 1348 and1350. Mary Matossian (1989) argues that the occurrence of the Black Death correlates with high humidity, rain, and flooding, which had characterized weather in much of Europe for two years prior to 1348, especially in England where the outbreak began. Europeans subsisted heavily on grains, and their trading and storage strategies also contributed to the spread of disease. The plague first ravaged commercial areas where stored grain surpluses attracted rats. When rats eat water-soaked grain they are likely to ingest *Fusarium* toxins and die, leaving their fleas to seek humans as new hosts.

However, medieval worldviews interpreted the demographic disaster as a divine punishment for nonrepentant wickedness. Churches elaborately frescoed to honor God were re-evaluated as a product of human arrogance and whitewashed. An entire religious order developed in which the central distinguishing ritual involved flagellating one's back to ribbons in hopes of obtaining pardon and ending the plague. Others celebrated the Black Mass, trying to appease the devil instead. When Europeans first traveled to the Americas, small pox traveled with them and decimated Native American communities. These likewise sought divine forgiveness and many among the Mandan, Omaha, and Cherokee killed themselves, and sometimes their families, to escape from torment to an afterlife.

One explanation for environmental crisis and the decline of the Easter Island population relates to the ceremonial erection of the multiton "ahu," or stone heads, for which the island is so famous. The vast amount of timber required for their transport and erection is thought to have depleted the island's forest resources. Soil depletion accompanied deforestation and led to decreasing crop yields. Clive Ponting (1991) suggests that in their anxiety about environmental changes, the Easter Islanders ritually erected more and more statues until they lacked even the wood to build canoes to fish or leave their depleted island-world. Rather than adapt their subsistence strategies, the islanders responded with religious rituals to alleviate their anxieties, but in doing so, contributed to their cause.

Among other challenges, the greenhouse effect is to bring drier weather in continental centers (the breadbaskets of many) and more violent storms. A quick look at recent human religious responses to uncontrollable environmental problems can give us an idea of how people might rationalize and ritualize their experience of climate change. In 1998, after decades of communist repression of religious expression, Romanians revived both Christian and pagan rituals during a time of searing heat and drought. In villages around Bucharest, a custom said to predate Christianity involved a pregnant woman covered in leaves and dancing in the fields to summon the gods of rain. In another Romanian custom called *Caloianul*, young girls tearfully bury a male clay doll and call for the blessing of the rain gods. Some fundamentalist Muslims in Turkey proclaimed in the streets, and to international news media, that the 1999 earthquakes were Allah's punishment for "wanton" jubilation over a preceding solar eclipse. Scientists have had a hard time convincing the general public that current warming trends are human induced and will have far-reaching impacts. With examples like these, we must wonder how people will respond when they begin to notice the effects of climatic change and when they understand human responsibility. Will we see a new Order of Flagellants?

HUMAN RESPONSE TO CLIMATIC CHANGE:
HISTORICAL ANALOGUES AND FUTURE CHALLENGES

We cannot begin to imagine the way in which an increase of just a few degrees in mean temperature will change the cultures of the world, especially those that subsist on marginal lands crossing national borders. With a warmer climate, insects and disease-bearing parasites will be able to spread to new environments. For centuries, the Fulani peoples of western Sudan have led their herds of zebu cattle to areas that they gradually learned were less conducive to the health of the tsetse fly (a bloodsucking fly that transmits sleeping sickness and the often lethal disease nagana to humans and livestock). Should the fly's habitat increase, the Fulani subsistence pattern and culture would certainly alter. This is only one example. Many traditional ways of life with subsistence strategies that gradually evolved for particular environments will have to adapt or die. Culture change should not be mourned for its own sake. Cultures are always evolving, disappearing, and emerging. However, the impact of global warming as currently predicted will result in such rapid environmental and climatic changes that cultures may respond too slowly to avoid human misery.

Although we may lack the analytical equivalent of a crystal ball to see how global warming will impact even the minutiae of our lives, we can examine, and learn from, responses to climatic change in the past through **historical ecology**. Carole Crumley (1994) suggests that the Romans could not have pushed their empire so far north had it not been for what is called the **Roman Climatic Optimum** (300 B.C. to 300 A.D.). This period was accompanied by a more stable and markedly warmer and drier climate. Mediterranean crops including grains and grapes could be grown even in Britain, so it behooved the Romans to acquire new suppliers (and taxpayers). Also spreading to northern Europe were the types of farming, patterns of settlement, land use, and even systems of land inheritance adapted for a mediterraneanized climate. This period of climate change forever altered Celtic culture. Romanization replaced the Celtic system of multiple-species agriculture and nonurban, dispersed settlement patterns. Crumley argues that the northward move of the Mediterranean-Temperate ecotone helps explain Roman conquest of the Celts. The return of a cooler climate also correlates with the retreat of Roman power.

Tom McGovern (1994) tells us of another climatic optimum that enabled the expansion of a people and their culture and how its end also led to their decline. During the **Medieval Climatic Optimum** (900–1200 A.D.) the Norse traveled to North America and successfully settled in Iceland and Greenland. They brought the methods of subsistence they employed in their homeland, but these were tenable only while the climatic conditions remained stable. Temperatures during the Optimum averaged 1° to 2°C (1.8° to 3.6°F) above the 1930–60 modern baseline, but cooled to 2° or 3°C (3.6° or 5.4°F) below this baseline in the mid-thirteenth century. Increasing drift ice made shipping more hazardous and isolated the Norse Greenlanders from Iceland and Scandinavia. Longer winters decreased the growing season and stunted pasturage for cattle. As Columbus was discovering North America again, the Norse Greenlanders were dying out.

McGovern reminds us that their extinction is not as simple as "it got cold and they died." Inuit Greenlanders prospered in the same time period by employing native subsistence strategies that the Norse were culturally loathe to adopt. The Inuit also moved about exploiting resources where they could, which the Norse, attached to their settlements and notions of personal property, would not do. In choosing not to innovate, the Norse Greenlander's demise was caused by culture rather than nature. Like the Easter Islanders erecting their stone heads, the Norse Greenlanders imported (when they could import anything) stained glass, expensive

Cultural Paradigms and Challenges

church vestments, and bells for the many large stone churches built in Greenland during the thirteenth century. They expended their remaining resources and energies on cultural symbols to alleviate their troubles, but which in the long run exacerbated them.

When we view these past human dilemmas, it seems obvious that a few degrees will make a difference in how we live and that our cultural visions of desirable or appropriate lifestyles will have to evolve. While the greenhouse effect may cause traditional societies to rapidly adopt new ways of living, industrialized societies may have to return to practices abandoned four to five generations ago. Only in the second half of the twentieth century did sprinklers and rubber hoses replace a centuries-old system of irrigation canals in the European Alps. As glaciers melt, the runoff must be efficiently channeled to avoid landslides. Italy's South Tirol and Austria's bordering North Tirol have already experienced devastating mudslides from meltwater. Many of the irrigation canals used and repaired communally by the great-grandparents of today's Alpine inhabitants have been destroyed or lost. Coordinated group efforts for their adaptive reuse may prove the best form of water control for guiding water across the paths of avalanches and over ravines. Meltwater will also have to be diverted around the multitude of lucrative new guest houses and ski resorts precariously perched on slide zones that farmers wisely avoided for millennia.

What of seasonal tourism? If a warmer climate means less snow in the areas that base their winter economies on skiing and snowmobiling, these areas might shift their tourist economies to summer. If they cannot guarantee snow, at least they can claim to be cooler than southerly areas in the warmest months. Sunny Florida, the Mediterranean, the Bahamas, and Morocco, warm-weather refuges for northerly dwellers, may lose their seasonal inhabitants and their appeal for retirees. Settlement and residential patterns, as well as the type of housing developed, may alter worldwide. Luxury developments on coastal areas and islands may lose value and structural integrity if we do indeed experience more floods and rising sea levels. Netherlanders are famed for their skill in engineering dikes to claim land from the sea, but with rising sea levels, will cities such as Amsterdam become like Venice? Some of the best underwater archaeology in the world is done around Denmark and Scotland, land areas that were joined about 10,000 years ago before the rapid global warming, glacial melting, and rising seas of the early Holocene. Sea levels in northern Europe did not reach current levels until almost 6000 B.C. by which time they had inundated the territories of a **Mesolithic** culture called Maglemosian. We may see many areas go the way of the mythical Atlantis, leaving much for future archaeologists to discover and much loss and perhaps tragedy in the process.

When we think of the many "if's" involved with global warming, we must remember to consider scale, both in our analysis and in discussion of impacts. Climatic factors shape human societies and cultures at every scale. How many people can live in one area together and for how long is directly tied to climatic conditions. Inuit bands dispersed in winter when resources were harder to find and, in warmer weather, when resources could sustain larger groups together, they gathered for sociability, marriages, and trade. Rainfall levels annually affect the way of life and social configurations of the !Kung of Botswana and the Aboriginal Australians. Beyond the local scale of bands, tribes, and chiefdoms, mini–ice ages have impacted settlement and subsistence strategies for entire regions of the earth. The greenhouse effect now confronts us with a human problem on the global scale. How will global climatic change impact the ways in which we live together in cities, villages, and rural areas? Which type of population aggregate will be most sustainable?

How we approach these challenges depends on finding common ground in our cultural perceptions of the environment. In building a model through which we can integrate our various cultural visions of the environment and the earth, we will have to work toward resolving false dichotomies between what is human and what is natural. In the West and Middle East, returning to a vision of ourselves as a part of nature rather than detached from nature is no easy project. Centuries of enculturation have taught us otherwise. To check your own perceptions, ask yourself a seemingly easy question: "Does my home shelter me from the environment or is the environment my home?" *Oikos,* the Greek word that gives us

Cultural Paradigms and Challenges   **97**

"ecology" means "habitation." Zen Buddhist architecture is designed to integrate a house within its natural surroundings so that it both encloses humans (provides shelter) and admits nature. To build a common vision and promote cooperative action, we must all realize that the environment is our common home, we are all Easter Islanders.

Conclusion

We too are on an island, but rather than collectively place our hope for environmental health in stone statues, we place it in science and technology. Technology for a time has outpaced the predictions of Thomas Malthus (that population increases will lead to global disaster). We have a hope-inspiring paradigm that science can solve current environmental challenges posed by global climate change. In the long run, however, this hope may also be our downfall; it certainly is a hindrance in convincing people, even those who believe global warming is anthropogenic, that we have a real problem. With our cultural mindset it is perhaps harder to believe that the environment is not inexhaustible, nor may be our scientific means to postpone or repair its depletion. Those who advocate a continuing reliance on the technological fix, for example, geoengineering, partially do so because they fear losing (or fear appearing to suggest that we change) a standard of living that is only a few generations old. Perhaps we should take a tip from the Norse Greenlanders and instead focus on the flaws in our standards of living to come to terms with its nonsustainability.

Our way of life derives from long-term sacrifices of environmental resources for short-term lifestyle gains. Making peace with the problems bequeathed to us may require short-term cultural sacrifices for long-term environmental gains. However, we need not think we must completely forsake life as we know it. Many answers are already with us if we just look for them. We can borrow ideas from other cultures, from the distant past, or from our grandparents' generation. We can take a dialectical approach to building on knowledge that has worked in the past and improving it to meet new challenges. The dialectic commonsensibly means shedding the strategies that no longer work, preserving those that do, and transcending them interdisciplinarily with new and better ideas. Our policies for coping with climatic change need not solely consider accommodation, adaptation, prevention, or geoengineering, but a carefully conceived combination of all approaches.

To be effective, our ideas must draw support from those of different cultural perspectives, socioeconomic classes, and educational and religious backgrounds. Instilling environmentalism into the diversity of cultural worldviews we find even within one society will require inspiration, creativity, and gushing goodwill beyond "hippie idealism." Controls on greenhouse gas emissions and the research and development of alternative forms of carbon-free energy may be costly in the short run. Realistic policies must be phased into society, so that individuals will want to integrate them into their lifestyles. When this is not possible, according to social or temporal pressures, legal coercion may be required. However, to be successful, this also must advocate constructive changes in such a way that it does not violate cultural beliefs in individual autonomy and freedom. Contrary to what advertisers would have us believe, the market is not the driving force behind culture. Raising prices for consumption of fossil fuels will not necessarily decrease their inefficient use, especially by the wealthy. To suggest policies paralleling "polluter pays" for global warming issues is difficult when skeptics see any such policy as "punishment," or as tied to political agendas, or feel someone (generally the government) is forcing a certain lifestyle on them.

When ethnic conflicts still destroy lives and land around the world, it is highly doubtful that any one policy can be accepted globally. In instituting constructive change, issues of policy implementation will be issues of scale. We will have to focus on regional and local policies and on a time scale adjusted to the needs and abilities of each area. Decision makers will have to sort out the responsibilities of markets, governments, and individuals for each area in relation to cultural, religious, and political traditions. Why go to such trouble? Why conduct fieldwork and historical research to create policies when people should just do what is right? Because what the evidence from our common human past shows us is that people do not automatically choose the right path or "do the right thing." If we are to, we must plan with equity and an awareness of the different needs, wants, and possibilities within regions, within nations, and within communities. What a challenge we have before us all.

REFERENCES

Crumley, C., ed. 1994. *Historical Ecology*. Santa Fe, New Mexico: School of American Research Press.

Ellen, R. 1982. *Environment, subsistence and system: The ecology of small-scale social formations*. Cambridge: Cambridge University Press.

Fagan, B. 1999. *Floods, famines, and emperors: El Niño and the fate of civilizations*. New York: Basic Books.

Ferkiss, V. 1993. *Nature, technology, and society: Cultural roots of the current environmental crisis*. New York: New York University Press.

Harre, R., J. Brockmeier, and P. Muhlhausler. 1999. *Greenspeak: A study of environmental discourse*. London: Sage Publications.

Harrison, G.G., W. L. Rathje and W. W. Hughes. 1992. Food waste behavior in an urban population. In *Applying anthropology: An introductory reader*, edited by A. Podolefsky and P. J. Brown, 99–104. Mountain View, Calif.: Mayfield.

Mannion, A. M. 1991. *Global environmental change: A natural and cultural environmental history*. Essex, England: Longman Scientific Technical.

Matossian, M. K. 1989. *Poisons of the past: Molds, epidemics and history*. New Haven, Conn.: Yale University Press.

McCay, B., and J. Acheson, eds. 1990. *The question of the commons: The culture and ecology of communal resources*. Tucson: University of Arizona Press.

McGovern, T. 1994. Management for extinction in Norse Greenland. In *Historical Ecology*, edited by C. Crumley, 127–154. Santa Fe, N.M.: School of American Research Press.

Milton, K. 1996. *Environmentalism and cultural theory: Exploring the role of anthropology in environmental discourse*. London: Routledge.

Ponting, C. 1991. *A green history of the world*. New York: St. Martin's Press.

SUGGESTED READINGS

Chambers, F. M., ed. 1993. *Climate change and human impact on the landscape: Studies in paleoecology and environmental archaeology.* London: Chapman and Hall.

Dincauze, D. 2000. *Environmental archaeology: Principles and practice.* Cambridge: Cambridge University Press.

Ferkiss, V. 1993. *Nature, technology, and society: Cultural roots of the current environmental crisis.* New York: New York University Press.

Fleming, J. R. 1998. *Historical perspectives on climate change.* Oxford: Oxford University Press.

Johnston, B. R. 1994. *Who pays the price? The sociocultural context of environmental crisis.* Covelo, Calif.: Island Press.

Kempton, W., J. S. Boster, and J. A. Hartley, eds. 1995. *Environmental values in American culture.* Cambridge, Mass.: MIT Press.

Keys, D. 2000. *Catastrophe: An investigation into the origins of the modern world.* New York: Ballantine.

Kottak, C. 1999. The new ecological anthropology. *American Anthropologist* 101:23–35.

Ladurie, E. L. 1971. *Times of feast, times of famine: A history of climate since the year 1000.* Garden City, N.Y.: Doubleday.

Merchant, C. 1995. *Earthcare: Women and the environment.* New York: Routledge.

Milton, K. 1996. *Environmentalism and cultural theory: Exploring the role of anthropology in environmental discourse.* London: Routledge.

Nazarea, V., ed. 1999. *Ethnoecology: Situated knowledge/located lives.* Tucson: University of Arizona Press.

Oelschlaeger, M. 1991. *The idea of wilderness: From prehistory to the age of ecology.* New Haven, Conn.: Yale University Press.

Orlove, B. 1980. Ecological anthropology. *Annual Review of Anthropology* 9:235–73.

Peterson, D., and D. Johnson, eds. 1995. *Human ecology and climate change: People and resources in the far north.* New York: Taylor and Francis.

Ponting, C. 1991. *A green history of the world.* New York: St. Martin's Press.

Rayner, S., and E. Malone, eds. 1998. *Human choice and climate change, Vol. I: The societal framework; Vol. III: The tools for policy analysis.* Columbus, Ohio: Battelle Press.

Redman, C. 1999. *Human impact on ancient environments.* Tucson: University of Arizona Press.

Reitz, E., L. Newsom, S. Scudder, eds. 1996. *Case studies in environmental archaeology.* New York: Plenum Press.

CHAPTER
5

Money, Markets, and Priorities

An Economic
Perspective on
Climate Change

Eban Goodstein

When faced with a global environmental issue like climate change, economists ask three questions. First, how much (if any) should humanity reduce emissions of **carbon dioxide** and other **greenhouse gases**? Second, once we have set a greenhouse gas reduction goal, how can we achieve that target at the lowest possible cost? And third, given that global warming and efforts to fight it each impose a different set of costs on different countries, how can global cooperation to reduce emissions be facilitated?

This chapter provides an introduction to how economists answer these questions. On some points, there is substantial agreement; on others, competing viewpoints still prevail. But while economists differ on some issues, we do agree on the centrality of these three questions.

101

HOW MUCH GLOBAL WARMING IS TOO MUCH?

At first blush, this may seem like a silly question. After all, global warming is likely to harm hundreds of millions of people around the planet, lead to far-reaching alterations in natural ecosystems, and dramatically accelerate species extinction. But an economic analysis of **global warming** (or any other environmental issue) begins with the realization that pollution is a by-product of material production and that people value both material goods and a clean environment. Given this, how should society balance the trade-offs?

In 1997, at a meeting in Kyoto, Japan, the industrial countries of the world signed the **Kyoto Protocol** (see Sooros, this volume, for further explanation of international legal agreements). This treaty, which has not yet been ratified by many countries including the United States, would require the signatories to reduce total emissions of greenhouse gases to 5% below 1990 levels by about the year 2010 (Goodstein 1999a). Is this the right goal? Should emissions be reduced further? Or not at all?

Economic Concepts

A UTILITARIAN FRAMEWORK

Economics as a discipline focuses on human welfare or well-being. For an economist, slowing global warming is valuable only because doing so increases the quality of life for people, those alive today and in the future. The impact of climate change on other species, independent of how humans are affected, is given no direct weight in the analysis.

This focus on **utilitarian,** anthropocentric concerns by economists is not meant to deny the importance of other ethical views. Indeed, broader moral considerations play a major role in the formation of social movements like environmentalism, and changing environmental norms will ultimately play a bigger role than a strict utilitarian calculus in shaping the condition of the planet that we pass on to our children and grandchildren.

Money, Markets, and Priorities   103

But economic arguments nevertheless loom large in most environmental policy debates.

In a utilitarian framework, **sustainability** can be defined in the following way. Sustainable outcomes require that we reduce greenhouse gas (GHG) emissions if this, on balance, prevents the decline of living standards below their current level for the typical (median) member of any future generation (Pezzey 1992).

MARKET SYSTEMS AND SUSTAINABILITY: EXTERNALITIES

Economists generally agree that unregulated market systems are likely to generate unsustainable outcomes in the face of climate change (Goodstein 1999b). In environmental terms, free-market systems fail because of the existence of negative **externalities**: costs generated in the production or consumption of a good that are not borne by the producer or the consumer of the good. When I buy a gallon of gasoline and burn it in my car, for example, I impose costs on society at large, ranging from the emission of local air pollutants to global warming gases (see Kay and Rall, this volume). Many of these social costs are not included in the price of gasoline-they are thus "external" to the buyer and the seller.

If all resources in an economy were privately owned, and environmental damages could be proved easily (with low "transactions costs"), then externalities would be internalized through private negotiation or litigation. For example, the Timberline Ski Lodge on Mount Hood in Oregon will (very likely) suffer future damage from my car's contribution to global warming, because snowfall is predicted to decline dramatically in the mountains of the Northwest over the next century as a consequence of climate change. As one possible solution to this problem, Timberline could sue me for damages and force me to "internalize" the external costs imposed on the ski lodge. However, there are millions of sources of greenhouse gas emissions across the planet, each inflicting a tiny cost on Timberline. Moreover, specific damages from any individual car would be very difficult to prove.

In the real world, of course, many important resources liable to be damaged by global warming are not privately owned—the air, rivers and streams, oceans, forests, deserts, and other natural habitats. No single party has a significant financial interest in protecting these resources. Given these two features—the high transactions costs associated with pressing court claims for damages and the existence of ubiquitous common property resources-economists agree that free-market systems will generate too much greenhouse gas pollution.

The conventional prescription for solving this problem is to require companies and consumers to internalize externalities by means of government regulation. This regulation can be prescriptive in nature, following a so-called command-and-control approach in which businesses are mandated to install specific types of cleanup technologies; alternatively, regulation can be incentive-based, including both pollution tax and marketable permit systems. These incentive-based approaches internalize externalities because they force businesses to pay for pollution, but they leave the particular method of reducing pollution up to the firms themselves (incentive-based approaches are discussed more later).

MARKET SYSTEMS AND SUSTAINABILITY: SHORT TIME HORIZONS

Although the externality problem can, in principle, be dealt with by means of regulation, free-market economies have a second feature that may generate unsustainable outcomes vis-à-vis climate change. In modern market systems, businesses require profit rates of around 20% in order to undertake investments in new technology. With such high rates of return, projects that generate payoffs that occur much more than five years in the future seldom appear attractive. Thus, from a business viewpoint, energy firms will show little interest in solar power as long as cheap oil and coal are likely to be available for at least another five years.

It is not obvious whether this short-term bias is unsustainable—that is, whether it actually *reduces* the quality of life for future generations below today's living standard. Even with a short-term bias, markets are very good

Money, Markets, and Priorities 105

at promoting the development of new technologies. It may be that a short market time horizon—coupled with government support for longer-term research—can still ensure a rising living standard for future generations, even as the climate changes. This will depend on the degree to which new technologies can in fact substitute for climate change–induced degradation of natural resources and **ecosystem services,** an issue addressed later in this essay (see also Herrod-Julius and McCarty, this volume).

HOW MUCH IS TOO MUCH?: THE BENEFIT-COST APPROACH

If the government forces firms to pay for the emission of global warming gases—"internalize externalities"—through regulation, how much "should" firms pay? The answer, from a benefit-cost (BC) perspective, is that firms should keep reducing pollution as long as the extra (or marginal) monetary benefit to society of the reduction of a ton of carbon is greater than the extra (or marginal) monetary cost. By following this rule, society will guarantee that the net monetary benefit of greenhouse gas reduction to society as a whole (total benefits minus total costs) is as great as possible.

Economists have developed several techniques to estimate the monetary benefits of reducing pollution, or put another way, the dollar value of pollution damages (Goodstein 1999b). These range from relatively simple impacts (e.g., fewer sick days for workers) to effects that are quite difficult to quantify in dollar terms (e.g., stroke deaths prevented, reduction in child IQ avoided, or biodiversity preserved). How do economists approach the latter set of issues?

In climate change benefit-cost studies, as one example, the dollar value of a life saved ranges from $2 million to $10 million. These estimates are based on the wage premium for risky employment in wealthy countries like the United States. Economists have found that controlling for other factors, workers such as police officers are paid a wage premium of close to $500 for accepting a 1-in-10,000 increase in the risk of death on the job. Given this, on average, 10,000 police officers wind up exchanging one of their lives for about $5 million.

Valuing life is one highly controversial part of **benefit-cost (BC) analysis.** A second is the process of **discounting.** Benefit-cost studies employ a formula in which benefits gained (as well as costs incurred) from future efforts to slow global warming are worth significantly less than benefits and costs occurring in the next few years. The rationale is that dollars spent on reducing carbon dioxide emissions is money that could be invested in alternative productive endeavors—developing new pharmaceuticals or building schools. Discounting helps guarantee that we do not spend "too much" on climate protection at the expense of other projects that will also generate dividends for future generations. Unfortunately, the conclusions of benefit-cost analyses are often sensitive to the choice of the discounting formula, and there is significant disagreement among economists about what the correct discount rate ought to be (Portney and Weyant 1999).

Benefit and cost uncertainties loom very large in the case of climate change. Analysts are required to estimate the dollar damages, 100 years hence, of CO_2 increases as they affect not only global agriculture and human health, but also species extinction and biodiversity. In view of these very large uncertainties, discounted economic estimates of the total global damages due to CO_2 doubling range from around negative 2% of global gross domestic product (GDP) to a positive 1% (an overall net benefit from warming). Across the planet, some regions will gain and others lose; impacts will be proportionately larger in poor countries and smaller in rich countries. Developing countries will be hit harder because they tend already to be in warmer and drier parts of the planet, but more important, because they have fewer financial resources for adapting their agriculture or building seawalls (Tol 1998).

Coupled with conventional, and again highly uncertain, estimates of the economic costs associated with reducing greenhouse gas emissions, a prominent group of benefit-cost modelers have argued for greenhouse gas emission reductions about 5% to 10% below a business-as-usual scenario. Note that this is not as strict as the Kyoto requirement of an emission freeze at 5% below 1990 levels. (However, when benefit-cost modelers employ low discount rates—and thus assume little reduction in

Money, Markets, and Priorities　　　**107**

the value of future benefits of cleanup—even these models call for cuts in emissions bigger than Kyoto, as large as 50% below a business-as-usual baseline [Weyant 1998; Kolstad 1998].)

Underlying the recommendations for modest emission cutbacks is a belief that climate stability is important, but not critical to the well-being of humanity. The argument proceeds like this: People adapt to changing resource conditions. As emissions of greenhouse gases are regulated, the global warming externality will be internalized. The price of CO_2-based services will then rise, and new, low CO_2 technologies will come online. Moreover, the development of new agricultural techniques will ensure that food supplies are maintained even in the face of a changing climate, and seawalls can be built to hold back rising sea levels. In addition, agriculture in some regions will gain from a warmer, wetter CO_2-enhanced climate and cold-related deaths will decline. Clearly, some people will lose from climate change, but on balance, the story goes, the quality of life for the median person in most countries will continue to rise even in the face of "moderate" climate change.

The benefit-cost perspective thus maintains that a policy of climate stabilization is too costly. Investing resources and person power in reducing greenhouse gas emissions will divert investment from schools or health care. Benefit-cost analysis is needed to obtain the right balance of investment between climate protection and other goods and services.

HOW MUCH IS TOO MUCH?: THE PRECAUTIONARY PRINCIPLE

In contrast to the benefit-cost approach, another group of economists see climate stability as a unique resource and have little faith that technological substitutes for it will be generated by real-world market systems. They maintain that for global warming, the future consequences of the current degradation of climate stability are too uncertain to justify the use of benefit-cost analysis. Instead, this group of economists argues for application of the **precautionary principle**. To preserve the welfare of future generations, therefore, climate stability should be protected *unless the costs of doing so are prohibitively high* (Daly 1996; DeCanio 1997).

Moreover, the argument is simultaneously advanced that the cost of protection will not be so high as is frequently claimed. Some analysts argue that already existing energy efficiency and renewable energy alternatives can get us from 60% to beyond the Kyoto reduction targets at low or no net cost (Laitner et al. 1998).

From a more philosophical perspective, advocates of a precautionary approach point to survey evidence suggesting that, in developed countries, broad growth in material consumption, in fact, leads to very little increase in societal well-being. Beyond a basic level, it is argued, human satisfaction from consumption depends on relative rather than absolute levels of consumption. To the extent that this is true, climate stability is being compromised only to support a "rat race" in which overall human welfare does not rise with increased consumption of SUVs and VCRs (Mishan 1968; Howarth 1996).

Which view is correct? Climate stabilization or modest reductions in greenhouse gas emissions? There is no "scientific" answer to this normative question, ultimately dependent on individual values. Nevertheless, both responses are "economic" in nature. Both agree on the common goal of sustainability presented above; both arise from a utilitarian philosophy, and both reflect a concern for trade-offs. Moreover, the debate is about ends, not means. When it comes to the latter, economists from both schools promote greater reliance on incentive systems than is provided by current government regulation.

Economics of Regulation

COMMAND-AND-CONTROL REGULATION

Since the early 1970s, many countries around the world have adopted national environmental protection legislation. Most of these laws mandated the development of highly detailed, prescriptive regulation for different types of industrial polluters. Economists have since labeled this **command-and-control (CAC) regulation.** CAC has two parts. The first is

Money, Markets, and Priorities **109**

uniform emission standards. All similar pollution sources are "commanded" to meet the same emission levels. The "control" portion of the name arises from the technology-based regulatory approach embodied in CAC. Regulators specify, in a very precise fashion, exactly which technologies companies must install to reduce emissions.

This initial wave of regulations achieved significant success. In the United States, for example, urban air pollution has improved significantly since 1970 and water quality has remained roughly constant, in spite of the fact that economic output has more than doubled (Goodstein 1999b). Nevertheless, economists have argued that CAC regulation is unnecessarily costly and that greater use of market incentives could cut pollution control costs significantly while achieving the same cleanup goals. Beyond that, some economists maintain that regulation by itself, whether CAC or market based, is inadequate for the task of achieving a sustainable economy. This group calls for greater government support for the development of a new generation of clean technologies.

It is easy to see how uniform standards and government-mandated technology requirements under CAC increase the costs of regulation. In the case of uniform standards, consider two neighboring power plants. Plant X, which has the ability to cheaply reduce carbon dioxide, is nevertheless required to meet the same standard as next-door power plant Y, a firm with very high costs of reduction. This is not a cost-effective solution: the same total emission reduction could be achieved for less money by having the low-cost firm meet a tighter emission standard, while easing the standard for the neighboring high-cost plant. Moreover, a single technological mandate is unlikely to provide the cheapest pollution control solution for different firms all over the country. The "one-size-fits-all" rigidity in centralized, technology-based regulation also increases costs (Goodstein 1999b).

However, the cost-raising effects of CAC in the long run are even more important. Technology-based regulation works against innovation on a couple of fronts. First, once the government-mandated technology is in place, a firm has no incentive to seek out ways to do better and reduce the

110 EBAN GOODSTEIN

remaining pollution that it emits. Moreover, should a firm want to develop an innovative, nonstandard approach, it must first obtain regulatory permission to do so.

INCENTIVE-BASED REGULATION

Incentive-based (IB) regulation is the alternative to CAC. There are two different types, which are closely related. The first is a **pollution tax** (also known as an effluent or emission charge or fee). For example, to reduce business emissions of carbon dioxide, which fuels the greenhouse effect (see Dobson and Kay and Rall, this volume), the government could institute a tax on emissions of $25 per ton of CO_2 released to the atmosphere. Alternatively, one might achieve a comparable reduction in emissions through a **marketable permit system** (also known as a tradable permit or cap-and-trade system). Under this approach, carbon-dioxide permits would be issued (or sold) to each polluter only up to a certain target level of emissions. These permits could then be bought and resold. Cap-and-trade systems, like emission fees, put a dollar-per-ton price tag on greenhouse gas pollutants.

Incentive-based systems eliminate uniform emission levels at different locations. Plants that face higher costs of reducing pollution will cut back only a little and either pay the emission fee or purchase additional permits. By contrast, firms with lower costs will cut back substantially and either save on pollution taxes or raise revenue by selling some of their permits. The identical systemwide level of pollution reduction can be achieved as under CAC regulation, but costs will be lower as companies exploit the greater flexibility and face incentives to make emission cuts at the lowest cost sites.

Another positive aspect of IB regulation is that it encourages innovation—firms can install any pollution control technology they desire. Moreover, IB regulation also sets up continuous incentives for technological progress. Under CAC regulation, firms are not penalized for any pollutants they emit once they have installed the technology mandated by the

Money, Markets, and Priorities 111

government. In sharp contrast, IB regulation requires firms to pay for every ton of greenhouse gas coming out of the stacks. The payment is direct under an emission fee system, but pollution generates a similar price under a cap-and-trade approach. As firms reduce pollution toward zero, they can make money by selling their extra permits.

An important caveat: For this system to work well, the government must have good monitoring and enforcement capabilities. Companies either must pay the emission fees that are due or, with cap-and-trade systems, face stiff fines for any pollutants they emit above their permitted levels. In some cases, command-and-control regulation may, in fact, be easier to monitor and enforce than an incentive-based system (Goodstein 1999b).

Emission fees as the sole instrument for controlling pollution are not often found in the real world, perhaps because as "new taxes" they are politically unpopular. By contrast, marketable permit systems are increasingly coming into use. The most well-known example to date has been the acid rain control initiative in the United States. Begun in 1995 to control the emission of sulfur dioxide (SO_2) from power plants, each utility receives annual SO_2 permits from the government equivalent to 30% to 50% of its 1986 pollution. Firms that come in under their permitted levels then sell permits; plants unable to comply purchase the excess permits. The program has functioned well. Emissions of SO_2 have fallen by about half, and the program has produced large cost savings (Schmalensee et al. 1998).

To attack the global warming problem, the Kyoto treaty includes the outline of a cap-and-trade system. If enough countries ratify the treaty and it goes into effect, then under the **joint implementation** provision the treaty will allow trading of carbon reduction credits between the developed countries. In addition, the so-called **clean development mechanism (CDM)** will allow rich countries to obtain carbon reduction credit for qualifying abatement activities that they finance in poor countries. There is some concern that trading systems may not work well in this case because of poor monitoring and enforcement.

PROMOTING CLEAN TECHNOLOGY

Regulation, whether CAC or IB, is needed in order to internalize externalities, forcing firms and consumers to bear the social costs of pollution.
Economists largely agree that a move toward greater use of IB regulation
would help lower the costs of environmental protection. But is "getting
the prices right," even through IB regulation, enough to ensure a sustainable reduction in greenhouse gas emissions?

Global warming emissions will be reduced substantially only if alternative (nonfossil fuel) energy technologies can offer competitive prices. Regulation raises the price of **fossil fuels** and thus encourages innovation. But
substantial price increases—the kind needed to really drive technological
innovation—are socially disruptive and thus politically difficult to institute.

Given these factors, a second approach is to promote directly the development and diffusion of clean technology: in the global warming case, fuel
cell vehicles, solar- and wind-powered electricity, energy efficiency, and
combined heat-and-power technologies. Rather than rely only on controlling greenhouse gases at the "end-of-the-pipe," the argument is that government should promote the use of technologies that reduce fossil fuel use in
the first place. These technologies can be advanced through policies such as
research and development subsidies, public procurement contracts, or tech-
nology-forcing regulation such as corporate average fuel economy (CAFE)
standards, energy efficiency standards, or the California ultra-low-emission
vehicle requirements. The former required car companies to improve
fleetwide fuel efficiency from 1975 to 1985; the latter mandates that 10% of
all cars sold in California by 2003 must be ultra-low-emission vehicles: fuel
cell, electric, or hybrid gasoline-electric (Goodstein 1999b).

Under a clean-technology policy, promotion should be restricted to
technologies that have the potential for medium-term commercial development, will generate a quality of service and have long-run production
costs comparable to existing technologies, *and* promise to substantially
reduce greenhouse gas emissions. The commercialization, quality, and cost
requirements must be met; otherwise, the new technology will not spread

Money, Markets, and Priorities   113

rapidly and little improvement in climate stability will result. But if these conditions hold, why aren't private firms developing these technologies fast enough in the first place?

The answer is found in the short-term bias in market behavior discussed above. Clean-energy technologies are unlikely to generate the kind of high short-run profits needed to attract substantial private investment. These technologies, unlike VCRs or cell phones, are not offering a new product; rather they must go head to head with existing, well-established technologies in mature industries—electricity or transportation. In these already competitive fields, only normal profits can be expected. Government policy can alter the profit equation by absorbing part of the risk of introducing these new technologies, nurturing infant industries that can then compete successfully.

Economists do not agree on how far government should move in this direction. Most economists believe that support of basic science is an important governmental role. Beyond that, some argue that government should promote large-scale investment in clean technologies, while others maintain that government involvement in technology policy should be minimal. Those opposed to government support for technology maintain that, if the "prices are right"—that is, externalities have been internalized—then the private marketplace will bring clean-energy technologies online soon enough to avoid unsustainable levels of environmental damage from climate change. Opponents also argue that the government may actually subsidize the wrong technologies and that technology decisions are better left to the marketplace.

Resolving Global Environmental Issues

Currently, the majority of greenhouse gas emissions are generated by developed countries (rich countries, countries in the "North"). However, within a couple of decades, developing countries (poor countries, countries in the "South") will surpass developed countries as emission sources, due to a combination of population and economic growth. Moreover, tropical deforestation is an important source of greenhouse gas emissions. Thus, global warming requires a truly global solution.

The 1997 Kyoto Protocol committed only the rich countries to reduce emissions below 1990 levels. The rationale here was twofold: First, greenhouse gases are long lived, so although poor countries will soon catch up in terms of annual emissions, the cumulative emissions from developed countries will remain primarily responsible for climate change over the medium term. Second, poor countries do not have the resources to invest in the alternative technologies—fuel cell vehicles, solar- and wind-powered electricity—that are needed to address the global warming problem.

The idea behind the Kyoto agreement was thus that rich countries would go first and develop low-cost alternatives to gasoline-powered automobiles and coal-fired power plants. These technologies would then spread from the North to the South, allowing poor countries to "leapfrog" the fossil fuel–based development patterns followed by the rich countries.

The Kyoto approach is modeled after the highly successful 1986 **Montreal Protocol**, an international agreement that mandated a phaseout of **ozone** destroying **chlorofluorocarbons (CFCs)**, used in the manufacture of air conditioners and refrigerators. Developing countries were given a ten-year grace period in the ozone treaty. Nevertheless, India and China initially refused to sign the Montreal Protocol, arguing that they did not want to sacrifice critical services for a problem that they did not cause. However, within a couple of years, cheap CFC replacements had been developed by companies in the North. In exchange for technological assistance in adapting their domestic industries to the new technologies, India and China signed on to the CFC production ban (Goodstein 1999b).

The Kyoto model as outlined above requires a way to transfer clean energy technology from rich countries to poor countries. The clean development mechanism under the current Kyoto accord—in which rich countries gain credit for clean investments in poor countries—is one prototype for technology transfer. If and when Kyoto is expanded to include limits on developing-country emissions, a global carbon cap-and-trade system will build on this approach. A trading system should provide excess carbon allowances to poor countries; these could then be "sold" to rich countries in exchange for access to solar power or fuel cell technologies or for help in protecting standing forests.

Basic economic theory suggests that international agreements like Kyoto are hard to develop. International pollution control agreements are **public goods**, goods that are enjoyed in common. The classic example, if a little dated, is the warning service provided by a lighthouse. Once in operation, it is impossible to exclude any passing boat from utilizing the warning beacon provided by the lighthouse. Public goods will generally be undersupplied in a pure free-market economy precisely because those who benefit can free ride if the good is provided at all by others. The existence of public goods provides one justification for government involvement in the economy: governments can use their ability to tax the public to provide an efficient level of public goods.

Reducing global warming is a public good. Free riders cannot be excluded from enjoying the climate stabilization benefits provided by an effective treaty. However, in this case, there is no *international* authority to ensure that national governments provide any climate stabilization effort. This public good nature of environmental treaties implies first that treaties will to be too weak from a benefit-cost point of view, because nations are reluctant to reveal their true willingness to pay at the bargaining table. Second, once a treaty is signed, each nation will have a strong incentive to fail to comply with the agreement. Unilateral cheating of this kind is a way to free ride on the pollution control efforts of others. Of course, if enough nations cheat, the agreement will fall apart.

With global warming, agreement on burden sharing has been difficult to achieve. Each country's willingness to pay depends on both the benefits received and the ability to pay. Both of these vary widely between nations. For example, low-lying Egypt has a very large stake in slowing global warming and its attendant, devastating sea-level rise. Yet the country is poor and will have a difficult time taking strong measures to reduce greenhouse gas emissions. On the other hand, a landlocked, wealthy country like Switzerland clearly has a high ability to pay but fewer interests at stake.

The willingness to pay on the part of a poor country to join an agreement will be much smaller than a rich country, simply because it has much lower national income. Yet, as noted, poor-country participation in greenhouse gas reduction will ultimately be essential. For example, if China further industrializes using its vast coal reserves, global warming will accelerate. But China has a low willingness to pay for a reduction in global warming, as a result of its low income. China will thus have to be compensated if it is to sign a greenhouse treaty. As noted, one way to induce China to join a treaty would be to provide it with excess carbon reduction credits that it can "sell" to countries in the North in exchange for nonfossil-fuel energy technologies.

The effort to implement the 1997 Kyoto global warming accord reflects both the difficulty in assigning burden sharing and the free rider phenomena. Strong opposition to treaty ratification has emerged in United States, led by President George W. Bush. This is not surprising, because the United States is the biggest greenhouse gas polluter. Bush and other treaty opponents argue that poor countries should be parties to the treaty and forced to reduce emissions on the same timetable as rich countries. This position threatens to derail the entire agreement (Goodstein 1999a).

The opposition to the Kyoto treaty in the United States is, in part, based on the belief that Kyoto would require large energy price increases and that U.S. manufacturing plants would flee to poor countries that are not part of the agreement. This argument has been made often in the past both by businesses opposed to environmental regulations and by environmentalists opposed to free-trade agreements. Economists have looked closely at this issue, and there is no past evidence for any significant capital flight on the part of firms seeking to escape environmental restrictions. Partly, this reflects the fact that environmental costs, even for heavily regulated businesses, rarely rise above 2% of total operating costs. Moreover, modern pollution control technology is often embedded in plant design. A new multinational chemical plant in Indonesia will look a lot like one in Kentucky. Although firms clearly do head South, the reason, overwhelmingly, is low wages (Goodstein 1999a).

Money, Markets, and Priorities   **117**

Given this record, regulation-induced **capital flight** should be considered a minor issue in global warming policy; there is, however, a more significant conflict between trade and climate change mitigation. The trading rules set up by the **World Trade Organization** (**WTO**) allow governments to challenge one another's regulations as trade barriers. For example, in 1999 the U.S. government, acting on behalf of Daimler-Chrysler and Ford, challenged a Japanese ruling designed to improve the efficiency of automobiles and reduce greenhouse gas emissions. The regulation required all midsized vehicles to meet a best-available-technology standard; in particular, engines had to be as efficient as one produced by Mitsubishi. The United States argued that the Japanese law discriminated against U.S. imports. If the U.S. challenge is upheld by the WTO tribunal, Japan will have to change its regulations or face the possibility of U.S. trade sanctions (Wallach and Sforza 1999).

Conclusion

This chapter has provided an overview of how economists approach the issue of climate change. Economists first ask what level of greenhouse gas emissions is best for the people who today, and who will in the future, inhabit the planet. Some economists answer this question in a benefit-cost framework; others, stressing uncertainty, advocate a precautionary approach.

Economists then ask how we can achieve our target CO_2 emission level at the lowest possible cost. Economists are in wide agreement that greater reliance on incentive-based approaches—emission fees or cap-and-trade systems—will save money. Beyond that, some economists advocate direct promotion of the development of clean technologies such as fuel cells and solar- or wind-powered electricity.

Finally, economists evaluate the role of international institutions. The theory of public goods suggests how difficult international agreements are to achieve; however, once in place, incentive mechanisms like a cap-and-trade system can both lower compliance costs and facilitate the transfer of clean technology from North to South. Other international goals, such as trade, need not necessarily conflict with climate stabilization policy, though institutions like the World Trade Organization may currently be working at cross purposes to national goals of greenhouse gas emission reductions.

How much warming is too much? How can we reduce emissions at low cost? And how can we design international institutions that align incentives with goals? An economic analysis of climate change helps shed light on the trade-offs inherent in policy choices. Where possible, economists try to discover and recommend policies that can minimize the scope of those trade-offs.

REFERENCES

Daly, H. 1996. *Beyond growth: The economics of sustainable development.* Boston: Beacon Press.

DeCanio, S. 1997. *The economics of climate change.* San Francisco: Redefining Progress.

Goodstein, E. 1999a. *The trade-off myth: Fact and fiction about jobs and the environment.* Washington, D.C.: Island Press.

———. 1999b. *Economics and the environment,* 2d ed. New York: John Wiley and Sons.

Howarth, R. B. 1996. Status effects and environmental externalities. *Ecological Economics* 16(1):25-34.

Kolstad, C. 1998. Integrated assessment modeling of climate change. In *Economic and Policy Issues in Climate Change,* edited by W. Nordhaus. Washington, D.C.: Resources for the Future.

Laitner, S., S. Bernow, and J. DeCicco. 1998. Employment and other macroeconomic benefits of an innovation-led climate strategy for the United States. *Energy Policy* 26(5):425-32.

Mishan, E. J. 1968. *The costs of economic growth.* London: Staples Press.

Pezzey, J. 1992. Sustainability: An interdisciplinary guide. *Environmental Values* 1(4):321-62.

Portney, P., and J. Weyant, eds. 1999. *Discounting and intergenerational equity.* Washington, D.C.: Resources for the Future.

Money, Markets, and Priorities

Schmalansee, R., P. Joskow, A. D. Ellerman, J. P. Montero, and E. M. Bailey. 1998. An interim evaluation of the sulfur dioxide trading program. *The Journal of Economic Perspectives* 12(3):53–68.

Tol, R. S. J. 1998. Comment on climate-change damages. In *Economic and policy Issues in climate change,* edited by W. Nordhaus. Washington, D.C.: Resources for the Future.

Wallach, L., and M. Sforza. 1999. *Whose trade organization? Corporate globalization and the erosion of democracy.* Washington, D.C. Public Citizen.

Weyant, J. 1998. The costs of carbon emission reductions. In *Economic and policy issues in climate change,* edited by W. Nordhaus. Washington, D.C.: Resources for the Future.

SUGGESTED READINGS

Cline, W. R. 1992. *The economics of global warming.* Washington, D.C.: Institute for International Economics.

Goodstein, E. 1999. *Economics and the environment.* New York: John Wiley and Sons.

———. 1999. *The trade-off myth: Fact and fiction about jobs and the environment.* Washington, D.C.: Island Press.

Koomey, J., R. C. Richey, S. Laitner, A. Sanstad, R. Markel, and C. Murray. 1998. *Technology and greenhouse gas emissions: An integrated scenario analysis using the LBNL-NEMS model.* Berkeley, Calif.: Lawrence Berkeley National Laboratory.

Lovins, A., and H. Lovins. 1997. *Climate: Making sense and making money.* Snowmass, Colo.: Rocky Mountain Institute.

Nordhaus, W., ed. 1999. *Economic and policy issues in climate change.* Washington, D.C.: Resources for the Future.

Portney, P., and J. Weyant., eds. 1999. *Discounting and intergenerational equity.* Washington, D.C.: Resources for the Future.

Romm, J. 1998. *Cool companies: How the best businesses boost profits and productivity by cutting greenhouse gas emissions.* Washington, D.C.: Island Press.

CHAPTER

6

Negotiating Our Climate

UNDERSTANDING
THE POLITICS OF
CLIMATE CHANGE

Marvin S. Soroos

One of the principal concerns of political science is to understand how governments address societal problems by adopting and implementing laws, policies, and programs, either on their own or in cooperation with other governments either nationally or internationally. Political scientists examine not only the actions of governments, but also the processes through which governing bodies make policies and the political forces that influence the actions of public officials. Governments will inevitably have to play a central role in efforts to address global climate change by regulating the human activities that are adding to atmospheric concentrations of greenhouse gases, such as carbon dioxide and methane.

Because climate change is a problem of truly global proportions, addressing the problem will require cooperation among the 200 or so nation-states that compose what is commonly referred to as the "international community." Under the principles of international law, these states claim to be **sovereign**, which means that they are not subject to a higher political authority that is empowered to impose rules on them without their consent. Thus, in the absence of a world government that can exercise such powers, it is up to this community of sovereign states to reach voluntary agreements on strategies for addressing problems of common concern. In doing so, states may voluntarily consent to rules or limits on their behavior and delegate responsibility for implementing such agreements to an existing or newly created international institution (Litfin 1998). These forms of "international governance," which include obligations and institutions that focus on a specific problem, are what political scientists call **international regimes** (Young 1994). Such regimes have evolved to address numerous international problems, such as managing use of the oceans and outer space, preserving endangered species, protecting the Antarctic environment, and reducing transboundary air pollution in Europe.

This chapter describes the international regime that is being created to limit global climate change. This regime is at an early stage of development compared with some of the other international environmental regimes, most notably the one that been constructed over the past fifteen years to preserve the **stratospheric ozone** layer. The ozone layer regime is arguably the greatest accomplishment thus far in international efforts to address global environmental problems. If the nations that have participated in its creation follow through on their commitments to phase out **chlorofluorocarbons (CFCs)** and other **ozone**-depleting substances, the ozone layer will begin within a decade to recover and may even be back to natural levels by the second half of the current century (Benedick 1998). The successful international response to the threat of ozone depletion has given rise to some optimism that climate change can be similarly constrained. Climate change, however, poses a much more formidable challenge for the international community, and thus it is not assured that another favorable outcome will result.

International Regimes

The term *international regime* has been used widely by political scientists since the 1970s to refer to the combination of existing norms, agreements, programs, and institutions that addresses a specific problem that is of concern to a grouping of nations or to the international community as a whole (Krasner 1983). This section reviews some of the typical components of regimes as background for discussing the climate change regime that has been developed over the past decade.

Nations that participate in international regimes acquire both privileges and obligations that take the form of international law. Some of these rights and responsibilities are based in traditional **international customary law**, which is norms of behavior that most nations have observed over time in comparable situations and are considered to be precepts of international law. Customary law is not written down in a formal sense but rather is determined by observing the behavior of states over time. Court cases decided by national or international tribunals and the writings of noted legal scholars offer interpretations of customary law. Customary law is considered to be mandatory for all countries, except for those that have consistently indicated their unwillingness to accept an emerging pattern of behavior as an international legal principle. A state's failure to disavow such a norm while it is gaining widespread acceptance among others is interpreted as implying consent to be bound by the norm (Malanczuk 1997).

The principles of international customary law are rather general and ambiguous guides for state behavior. In any given situation they may be susceptible to competing interpretations. Thus, such principles tend to be inadequate for determining what is expected of states to address complex global problems such as climate change. In recent decades, the international community has been supplementing international customary law with negotiated agreements, known as **treaties**, which are written documents that spell out in more specific language what is expected of the states that formally accept them. Treaties may apply to as few as two states or they may be products of negotiations open to all states, in which case they are commonly referred to as **conventions**.

The acceptance of treaties by states is a multiple-stage process. Once the representatives of the nations participating in treaty negotiations are satisfied with the wording of the agreement, they are given an opportunity to sign the agreement for their countries. Such a signature does not officially oblige a state to comply with the provisions of a treaty or a convention. It does, however, suggest that the branch of government participating in the negotiations will proceed to the next step in the treaty approval process, which is **ratification** of the agreement through appropriate constitutional processes of its country. In the United States treaties are ratified by a two-thirds vote in the Senate. Nations that have ratified a treaty are then required to comply with its provisions only after it "enters into force," which occurs once ratification papers have been received from a certain number of countries specified in the treaty (Janis 1999).

International bodies or specially convened conferences, such as those of the United Nations, often adopt another type of document usually called a **resolution**, but also may be referred to as a declaration, a statement of principles, or a plan of action. Typically, such documents set forth principles and guidelines for addressing problems, and in some cases spell out concrete measures that might be carried out. These documents may be adopted by a majority vote of the nations represented in the body or by a less formal consensus. Under most circumstances, such documents, including those adopted by the General Assembly of the United Nations, are merely recommendations and no country is legally obliged to comply with their recommendations, even the countries that voted for them. For this reason, resolutions and similar documents are sometimes described as "**soft law**," in contrast to customary law and treaties that are referred to as "hard law" because they are binding on at least some countries. Such documents are also sometimes called "instant law" because they can be debated and adopted rather quickly, in contrast to treaties that may take years or even decades to negotiate. Despite their marginal standing, resolutions or similar documents can be an important step in the evolution of international regimes, as they often provide a statement of principles that becomes a basis for more formal negotiations of a treaty (Jurgielewicz 1996).

International institutions, in particular the United Nations and affiliated organizations and programs, play an essential role in the creation and development of international regimes. In some cases, they sponsor and provide the venue for negotiations on international treaties. The United Nations Environment Programme (UNEP) has performed this role for negotiations on a number of major environmental treaties, including those designed to protect the ozone layer. International organizations, in particular the specialized agencies of the United Nations, coordinate and facilitate international scientific programs that monitor environmental trends and contribute to a deeper understanding of the dynamics of the earth system. The World Meteorological Organization and UNEP have jointly sponsored the World Climate Programme, which was created in 1980 to monitor and investigate the phenomena of climate changes, including those that occur naturally or are human induced. The same two organizations also oversee the Intergovernmental Panel on Climate Change (IPCC), which since it was formed in 1988 has engaged more than 2,000 scientists from around the world to sift through the findings of scientific studies that shed light on climate change and its possible impacts and to issue reports to inform negotiations on the problem. International organizations also play an important role in the implementation of treaties, such as by reviewing national reports on steps taken to comply with international regulations and by monitoring and publicizing violations (Werksman 1996).

THE GLOBAL CLIMATE CHANGE REGIME

International concern over the prospect of human-induced global warming and its potentially catastrophic impacts rose sharply during the 1980s in response to increasingly urgent warnings from the scientific community.

Moreover, a trend toward markedly warmer global average temperatures during the decade as well as a spate of unusual weather phenomena around the world seemed to bear out the forecasts of scientists. A series of major international conferences convened toward the end of the decade took up the subjects of climate change and what could be done to limit it to manageable proportions (see figure 1). The Canadian government sponsored the first of these meetings in Toronto in 1988 called the Changing Atmosphere Conference. The Intergovernmental Panel on Climate Change (IPCC) issued its first report in 1990, warning of a 1° to 3°C (1.8° to 5.4°F) increase in global average temperatures over the next century if steps were not taken to significantly limit emissions of **greenhouse gases (GHGs)** (IPCC 1990). Following up on the recommendations of the Second World Climate Conference, which was held in Geneva later in 1990, the United Nations General Assembly voted to convene an International Negotiating Committee (INC) to draft an international agreement on climate change (Soroos 1997).

The INC succeeded in negotiating a treaty known as the Framework Convention on Climate Change (FCCC), which was ready for adoption at the Earth Summit in Rio de Janeiro in 1992. A **framework convention** or treaty is an initial agreement that lays the foundation for negotiating additional measures, often referred to as **protocols,** as circumstances dictate. The FCCC is similar to framework agreements that address other environmental problems, such as transboundary air pollution and depletion of the ozone layer, in that it does not legally oblige the parties to reduce emissions of the pollutants that are responsible for the problem. Nevertheless, the FCCC is notable for several reasons. First, it sets forth an ambitious goal of "preventing dangerous anthropogenic interference" with the global climate system, which already appears to be an unrealistic objective. Second, it explicitly acknowledges that the industrialized countries are largely responsible for past human additions to concentrations of greenhouse gases in the atmosphere and calls on these countries to take the lead in combating climate change. Third, to demonstrate their commitment toward taking the first steps toward addressing the problem, the industrial countries, which are listed in Annex I of the treaty, agreed to a

1988	The United Nations (UN) organizes the Intergovernmental Panel on Climate Change (IPCC). The Changing Atmosphere Conference takes place in Toronto.
1990	The first IPCC report is released projecting a mean global temperature increase of 1° to 3° C (1.8° to 5.4° F) by the year 2100. The Second World Climate Conference is held in Geneva.
1991	The United Nations convenes an Intergovernmental Negotiating Committee (INC) to begin drafting the Framework Convention on Climate Change (FCCC).
1992	The FCCC is finalized and signed by more than 150 countries at the Earth Summit in Rio de Janeiro. The United States is among the first countries to ratify the treaty.
1995	The first Conference of the Parties (COP I) is held in Berlin. Parties adopt the Berlin Mandate, which commits them to conclude an agreement at COP III in 1997 that would include a timetable for binding emissions for the industrial countries. The IPCC releases its second report with warming projections from 1° to 3.5° C (1.8° to 6.3° F) by 2100.
1996	COP II is held in Geneva. Parties attending endorse the second report of the IPCC and continue planning for a future protocol to limit GHG emissions.
1997	COP III is held in Kyoto. More than 150 countries adopt the Kyoto Protocol establishing a schedule for emission limitations by the developed countries, but details remain to be worked out. The Clinton administration agrees to a 7% reduction in U.S. emissions from 1990 levels by 2012.
1998	COP IV is held in Buenos Aires. The United States insists that developing countries also commit to emission limits.
1999	COP V is held in Bonn. Numerous issues remain unresolved, but the parties commit to finalizing the Kyoto Protocol at COP VI scheduled for 2000.
2000	COP VI is held in The Hague. A stalemate between the United States and the European Union thwarts final agreement on the Kyoto Protocol.
2001	IPCC releases its third report projecting a warming of 1.4° to 5.8° C (2.5° to 10.4° F) by 2100. The Bush administration rejects the Kyoto Protocol. Final version of the protocol is adopted at a continuation of COP VI in Bonn without the concurrence of the United States.

Figure 1. Time line of climate change negotiations. (Not all dates included in this time line are discussed in the text.)

nonbinding goal of returning their emissions of greenhouse gases to 1990 levels by the year 2000. These so-called **Annex I countries** are also to provide reports detailing how they plan to reduce GHG emissions and enhance carbon sinks (Bodansky 1993).

Efforts to strengthen the international climate change regime have been undertaken at annual Conferences of the Parties (COPs). The first of these meetings, known as COP I, took place in 1995 in Berlin a year after the agreement came into force, after having been ratified by the requisite fifty states. It was readily apparent to the nations represented at COP I that the FCCC was by no means an adequate response to the climate change problem, but they were still unable to reach a consensus on mandatory limits on emissions of greenhouse gases. However, the delegates made a commitment to have an agreement ready for adoption at COP III in Kyoto in 1997, which would specify mandatory emission limits for the developed countries for years 2000, 2005, and 2010 (Rowlands 1995).

The prospects for a significant agreement did not appear favorable as the dates for COP III approached, nor was much progress apparent during the first ten days of the conference. Then, at an all-night session convened after the conference was scheduled to end, the delegates hammered out a document that has become known as the **Kyoto Protocol** of 1997. The most significant feature of the protocol is the schedule for mandatory limitations of GHG emissions by the developed Annex I countries. The limits pertain to a package of six greenhouse gases (carbon dioxide, methane, nitrous oxide, hydrofluorocarbons, perfluorocarbons, and sulphur hexafluoride).

In contrast to common practice in other environmental treaties of requiring the same percentage of reduction for all countries, known as "across-the-board" cutbacks, the Kyoto Protocol has "differentiated" reduction targets for the developed Annex I countries. These targets specify 1990 as the base year and the limits are to be achieved by the period 2008 to 2012, with an average figure being calculated for these five years. Most European countries, including the members of the European Union, agreed to reduce their GHG emissions by 8%, the United States by 7%, and Japan and Canada by 6%. The Russian Federation and Ukraine merely agreed to hold their emissions to 1990 levels. Several other states agreed only to restrain their increases, the most notable example being Australia, which would keep its emissions from rising more than 8%. The rationale for these varied limitation targets was that some countries had unusual circumstances that would make it prohibitively expensive, if not impossible, for them to achieve the more ambitious reduction targets of other countries. Collectively, these commitments would achieve a 5.2% reduction in GHG emissions of the developed Annex I countries. Nothing in the Kyoto Protocol obliges developing countries to reduce or even to limit their emissions.

The Kyoto Protocol is also notable for the inclusion of what have been called **flexibility mechanisms**, which offer the developed countries an array of options for achieving their limitations on GHGs emissions. One is the possibility of taking steps to sequester additional carbon as way of offsetting emissions of greenhouse gases. This might be done by planting trees to increase forest cover or by various agricultural practices, such as no-till cultivation. Another flexibility mechanism would institute a system of international emission trading in which countries that were not achieving their emission limitation target domestically could purchase emission credits from countries that more than complied with their obligations to limit emissions. The two countries most likely to have substantial credits to sell are the Russian Federation and Ukraine, which have seen a substantial drop in their GHG emissions during the 1990s, as inefficient, antiquated industries have been shut down and the economies of these countries have fallen into a deep recession. Such permits for emission reductions that would have occurred even without the protocol are critically referred to as "hot air," because they would undermine the purpose of the agreement.

There is the possibility for cooperative ventures known as **joint implementation (JI)**, in which Annex I countries would receive emission reduction units (ERUs) for investing in projects in other Annex I countries that would result in emission reductions in addition to those that would have been achieved otherwise in these countries. It is anticipated that it will be far less costly for the highly developed Annex I countries to acquire ERUs than to achieve an equivalent amount of reductions domestically. Annex I countries could use the ERUs that they have accumulated toward fulfilling their obligations to reduce or limit GHG emissions. The protocol also establishes a **clean development mechanism (CDM)** through which Annex I countries, or industries based in them, could invest in JI projects in developing countries that would provide them certified emission reductions (CERs). This could be done, for example, by investing in the development of renewable energy sources or simply by projects that would improve the efficiency of **fossil fuel** use. Another option would be to contribute to forestation programs that would sequester more **carbon dioxide**. The developed country or company making the investment would then receive credit for whatever net reductions in GHG emissions resulted, which could take the place of reductions they would otherwise be obliged to achieve domestically (Ott 1998).

Finally, the so-called bubble provision permits a group of countries to achieve a reduction target collectively. Under this arrangement, some countries in the group could be given less ambitious targets as long other countries were committed to achieving greater GHG reductions than the group's average. This flexibility mechanism was especially attractive to the European Union (EU) whose members vary considerably in their capacity to cut back on emissions. Thus, although the EU countries are obligated to reduce their emissions by 8%, Germany and Denmark agreed to reduce their emissions by 25% and the United Kingdom by 10%, whereas Greece and Portugal are allowed increases of 25% and 27%, respectively. If the EU countries fail to cut their collective emissions by the amount designated in the Kyoto Protocol, then each country is obliged to comply with the 8% reduction (Ringius 1999).

In the rush to conclude the Kyoto Protocol, the tired delegates postponed decisions on many of the technical details regarding how the agreement would be implemented and enforced. On many of these details, the conflicting positions of states have defied a quick compromise. Furthermore, how these issues are ultimately resolved will have a considerable impact on the effectiveness of the protocol in achieving its objectives. Thus, efforts to work out the remaining details of the agreement have proceeded at COP IV in Buenos Aires in 1998 and COP V in Bonn in 1999 (Lanchbery 1998). It was hoped that COP VI in The Hague in 2000 would finalize the Kyoto Protocol, but the talks collapsed in acrimony and disarray, as the European Union was firm in refusing to yield to a U.S. proposal that countries be allowed to achieve part of their emission reduction commitments by preserving existing carbon sinks, such as forests and range land.

Early in 2001, the new U.S. president, George W. Bush, summarily rejected the Kyoto Protocol, arguing that the agreement was "flawed" and would damage the nation's economy. Despite this major blow to the Kyoto process, COP VI was reconvened in Bonn in July 2001. With the United States on the sidelines during the negotiations, the other 184 nations in attendance worked out the remaining details of the Kyoto Protocol.

The Politics of Climate Change

INTERNATIONAL POLITICS

Treaties are the product of negotiations between representatives of nations in which there is often a high level of disagreement and conflict among blocs of countries. Negotiators tend to seek an agreement that maximizes the benefits that their states will derive from it, while minimizing the obligations and sacrifices that are required of them. Thus, the process of negotiating major treaties can drag on for years, and in some cases even over a decade or more, during which a series of negotiating sessions is held, each of which lasts a few weeks. Because states cannot be compelled to ratify treaties and be bound by their provisions, an effort is made to draft language that all states can accept, in particular the larger, more influential ones whose refusal to ratify a treaty could render it ineffective. The task of reaching a consensus on a major treaty is especially daunting when the negotiations are open to all states, which has been the case with the efforts to forge treaties to limit global climate change (Muldoon et al. 1999).

Three major groupings of countries have shaped the negotiations on limiting global climate change: (1) a coalition composed largely of European countries, which has consistently pushed for strong international agreements that set target dates for binding reductions of greenhouse emissions, (2) a small bloc of the remaining developed countries, led by the United States, that has resisted committing to mandatory reductions of GHG emissions, and (3) a large bloc of developing countries that argues the industrialized countries are responsible for the problem and should take the first steps to address it. Several other groups also played active roles in the negotiations, including the countries-in-transition of the former Soviet bloc, whose GHG emissions have declined sharply over the 1990s due to their economic restructuring; an alliance of small island nations that is especially vulnerable to rises in sea levels associated with global warming; and a group of oil-producing countries concerned that agreements designed to reduce the use of fossil fuels would significantly reduce their export revenues.

Even before official negotiations on the framework convention commenced in 1991, most industrialized countries, the notable exception being the United States, were calling for a schedule for cutting back GHG emissions. Several states even made unilateral commitments to stabilize if not substantially reduce their emissions in hopes of setting an example for other countries. These states argued strongly that mandatory emission reductions by the developed countries should be included in the FCCC being prepared for adoption at the Earth Summit. They were disappointed by the refusal of the United States to agree at the time to binding reductions, but in the interests of keeping the United States in the process, compromised on the voluntary goal of stabilizing GHG emissions at 1990 levels by 2000. As negotiations resumed at the COPs beginning in 1995, most of these countries to continued to push for a firm schedule for reducing emissions (Soroos 1997). As the Kyoto COP III approached, the European Union proposed a 15% reduction in emissions by 2010.

Negotiating Our Climate 133

The European-based coalition has been joined by a group of countries that has formed the Alliance of Small Island States (AOSIS) in 1990. This alliance, which has grown to forty-three states that consist of islands or low-lying areas from all oceans and regions, has consistently been making the most radical proposals for mandatory GHG emission reductions by the developed countries. Prior to the Kyoto meetings in 1997, the AOSIS was calling for a 25% cutback by 2005. While the AOSIS countries are small and even tiny by most measures-territory, economy, and population-the thirty-six of them that are members of the United Nations account for 19% of the votes in that organization. The leaders of small of the smallest and most vulnerable of these countries, such Kiribati, Tuvalu, Vanuatu, and the Maldives, have repeatedly appealed to the conscience of the world community, maintaining that their existence and the survival of their cultures are being threatened by a problem caused largely by the industrial countries (Acharya 1995).

At the time the FCCC was negotiated in the early 1990s, the United States was virtually alone in its opposition to a schedule for mandatory reductions of emissions of GHGs. Its only consistent allies in the talks were the oil-producing countries that had a vested economic interest in not limiting the consumption of fossil fuels. The United States tried to justify its position by noting that the science of climate change and its impacts were still too uncertain to warrant postponing limits on emissions of GHGs that could prove very costly to implement. The United States was receptive to adopting so-called no regrets policies to address climate change, which could be justified by other benefits if the threat of global warming proved to be less urgent. The United States also argued that any reduction scheme should not be limited to carbon dioxide, but should be comprehensive in including other known GHGs, such as chlorofluorocarbons (CFCs), which were being phased out by the **Montreal Protocol** and its later amendments.

As the Clinton administration took office in 1993, hopes ran high that the United States would be more willing than its predecessor to accept binding reductions of GHG emissions. These expectations were not realized, at least initially, as the new administration was also hesitant about making such a commitment. Moreover, the United States was joined in the negotiations by several other countries that had reservations about agreeing to substantial reductions of GHG emissions, including Australia, Japan, Canada, New Zealand, and Norway, which along with the United States became known as the Umbrella Group. As the date for COP III in Kyoto approached, the United States set forth a plan that would include $6 billion in investments in research on energy efficiency and alternative energy sources over the next decade and an effort to stabilize GHG emissions at 1990 levels by 2008 to 2012. This proposal was strongly criticized by the European countries and environmental groups for being much too weak.

The United States surprised most observers when it agreed at the Kyoto COP to reduce its GHG emissions by 7% from 1990 levels by 2008 to 2012. The acceptance of this goal, however, was contingent on the inclusion of "flexibility" mechanisms, in particular emission trading and joint implementation, which would provide the United States, and other Annex I countries, a less costly option of achieving much of its reduction in other countries. The alternative of sequestering carbon, such as by expanding forest cover, also made the agreement more palatable to the United States as did the five-year target period over which emissions could be averaged. Other Umbrella Group countries took advantage of the opportunity for differentiated limitations on GHG emissions to avoid cutbacks or to merely agree to restrict their increases. Australia, which had previously joined European countries in supporting emission reductions, had a change of heart under its new conservative government and argued that its unique circumstances justified an 8% increase in emissions (Christoff 1998).

The COP VI negotiations broke down when the European Union refused to yield to the United States on several key issues. Among these was a U.S. proposal that countries be permitted to claim credit for preserving existing sinks, such as forests, cropland, and rangeland to partially fulfill their emission reduction targets. The United States also insisted that no limits should be placed on emission trading as an option for meeting its Kyoto obligations. Representatives of leading European countries, such as German and France, were concerned that permitting so much flexibility for achieving emission targets would substantially diminish whatever

Negotiating Our Climate

impact the Kyoto Protocol might have on limiting the flow of GHGs into the atmosphere. They were also adamant that the United States should achieve a large portion of its commitment to reduce GHGs domestically by curbing its excessive consumption of fossil fuels.

The developing countries make up by far the largest bloc in the climate change negotiations. As in many other negotiations, these countries meet and coordinate their negotiating positions through a coalition of Asian, African, and Latin American countries known as the Group of 77/China. The Group of 77/China began caucusing together in United Nations forums beginning in 1964 and has grown to 133 members (Williams 1997). Throughout the negotiations on the climate change regime, the Group of 77/China has emphatically argued that the industrialized countries are primarily responsible for human additions to GHGs to the atmosphere. Accordingly, fairness would seem to dictate that these countries act first to cut back their emissions, and thus to mitigate climate changes that could impact more heavily on developing countries. The Group of 77/China has been nearly unanimous in rejecting any proposals for mandatory limits on their GHG emissions, which they contend would hamper their legitimate aspirations for economic development and reducing poverty. Nor have the developing countries been very supportive of the flexibility mechanisms written into the Kyoto Protocol on grounds that the developed countries, and in particular the United States, should not be able to use their wealth to avoid making significant sacrifices to reduce their GHG emissions (Dunn 1998).

Figures on emissions of carbon lend considerable legitimacy to the position taken by the Group of 77/China. The industrialized countries, including those of the former Soviet bloc, account for approximately 75% of cumulative carbon emissions since 1950. Calculated on a per capita basis, the carbon emissions of the developing countries continue to be a small fraction of comparable figures for the industrial countries. China's per capita emissions are only about one-eighth of those of the United States, whereas those of India are only about one-twentieth the U.S. figure. From the perspective of the developing countries, it is highly unjust that the industrialized countries are the principal contributors to climate changes that are likely to have more severe impact on developing countries (Engelman 1995). Moreover, Third World countries have far fewer resources with which to adapt to the unwanted consequences of climate change.

Most of the developed countries acknowledge the legitimacy of the position of Group of 77/China on emission reductions. The United States, however, has been insisting in recent years that it will ratify the Kyoto Protocol only when the developing countries agree to constrain their GHG emissions. The United States notes that the emissions of developing countries are rising more rapidly given their economies and populations. Developing countries currently account for approximately 40% of the world's carbon emissions and by 2020 are predicted to surpass the annual emissions of the industrialized countries. By 2015 China, with an economic boom fueled by its vast coal reserves, may surpass the United States as the world's leading emitter of carbon (Dunn 1998). The United States also expresses the concern that industries using large amounts of fossil fuels may relocate to developing countries that impose no limits on GHG emissions. For these reasons, the United States argues that any cutbacks in GHG emissions by the developed countries may be nullified by rising emissions in the developing countries.

The countries in transition that composed the former Soviet bloc have played a less prominent role in the climate change negotiations than the blocs mentioned thus far. These countries have faced difficult challenges in the transformations to democratic forms of government and market-based economies. Thus, environmental problems, especially global ones, have not been high on their list of priorities during the 1990s. The restructuring of their economies has led to the closing down of numerous antiquated industrial plants that were highly inefficient in their use of fossil fuels and thus major emitters of carbon. The closing of these plants along with the steep economic declines in Russia, Ukraine, and several of the other countries in transition has resulted in drops of 30% or more in GHG emissions during the 1990s. Both Russia and Ukraine agreed to zero net change in GHG emissions from 1990 levels under the Kyoto Protocol, which offers the prospect of earning billions of dollars through the sale of emission permits to other industrial countries, such as the United States.

Thus, these countries have been generally supportive of the Kyoto Protocol and its flexibility mechanisms.

U.S. DOMESTIC POLITICS

Political scientists are not content simply to describe the international regimes that have been created to address a given problem. They also seek to understand *why* the process resulted in a given outcome, be it a success or failure. Explanations of why states take the positions they do in negotiating or ratifying international climate change agreements typically consider factors such as size of the country, level of economic development, endowment of natural resources, dependence on fossil fuels for energy, costliness of reducing GHG emissions, and vulnerability to climate changes and their impacts. Of particular interest is the politics of climate change within countries, as shaped by their governing institutions, perceptions and priorities of leaders and policy makers, influences of nongovernmental groups, public opinion, and general political cultures. The internal politics of climate change defined by these variables is unique to each country. This section will focus upon the politics of climate change within the United States in view of its disproportionate share of global GHG emissions and the importance of its role in international efforts to address the problem.

The political system of the United States is distinctive in two ways. First, there is a well-defined division of power among the executive, legislative, and judicial branches of government. Thus, the Congress operates with an independence of the presidential branch that is in marked contrast to most European countries in which prime ministers work closely with the parliamentary majorities that put them in office. The relationship between the U.S. president and Congress is often contentious, especially when there is a divided government, with one party controlling the presidency and the other the houses of Congress, as has been the case during all but one legislative session over the past two decades. The U.S. government system is also notable for having a federal structure in which there is a division of political authority between the national government in Washington, D.C., and the governments of the fifty states.

Negotiations on the climate change regime have spanned the elder Bush, Clinton, and younger Bush presidencies. The elder Bush administration strongly resisted committing to a firm schedule for cutting back on GHG emissions, which frustrated efforts by other industrialized countries to write mandatory reductions into the FCCC. While numerous federal departments and agencies have been involved in formulating the U.S. climate change policy, it was widely believed that President Bush's position was strongly influenced by his chief of staff, John Sununu, who was openly skeptical about the scientific evidence of human-induced climate change. The 1992 election of the Democratic ticket of Bill Clinton and Al Gore, the latter an outspoken believer in the threat of climate change, raised hopes of a major shift in American policy toward supporting a stronger international climate change regime (Gore 1992). Although the Clinton administration finally acceded to a 7% reduction in GHG emissions under terms of the Kyoto Protocol, environmentalists were dismayed at the administration's procrastination in making such a commitment and its failure to do more to mobilize the public support needed to prod the Senate to ratify the protocol. The younger Bush administration, with its close ties to the petroleum and other energy industries, was quick to disassociate itself from the Kyoto Protocol that the Clinton administration negotiated with 150 countries at COP III. Thus far, the Bush administration has not offered any alternative approaches for addressing climate change, while standing back to watch most other countries work out the final details of the protocol.

Congress plays two important roles in U.S. climate change policy. First, ratification of treaties negotiated by the presidential branch of government is the constitutional responsibility of the Senate. Treaties must have broad bipartisan support if they are to receive the two-thirds vote of senators required for ratification. A ratified treaty becomes the law of the land in the United States, which is not necessarily the case in other countries. Second, the House and the Senate must then jointly adopt legislation needed to bring the United States into compliance with whatever obligations the country has under terms of the agreement. Congress tends to take a skeptical view of the treaties negotiated by the presidential branch. Thus, numerous treaties in realms such as arms control, human rights, and the environment have gone unratified, even after most other countries have formally accepted them. A failure of the Senate to ratify a treaty does not, however, preclude Congress from taking steps to implement the provisions it finds acceptable.

In recent years, the Republican-controlled Senate has taken a very critical view of ongoing international efforts to address climate change, in particular the Kyoto Protocol. Several months before COP III in Kyoto, the Senate adopted a resolution by a vote of 95-0 that counseled the Clinton administration not to sign an agreement mandating new reductions or limits of GHG emissions for the United States that (a) did not also require a similar commitment from developing countries to limit or reduce their emissions or (b) would result in "serious harm to the economy of the United States." While the climate change issue has increasingly taken on a partisan tone, with Democrats generally expressing more concern about the threat of global warming, the unanimous vote of the Senate on its pre–Kyoto resolution revealed that many Democratic senators were also unwilling to support the anticipated type of protocol. Among the Democratic senators voting for the resolution were those from states that are major producers of fossil fuels or automobiles. The willingness of the administration to agree to a protocol requiring a 7% emission reduction by the United States with no similar requirement for developing countries seemed to defy the advice of the Senate, provoking some senators to declare the agreement would be "dead on arrival" if submitted to them for ratification. To avoid having the protocol resoundingly defeated, the Clinton administration withheld it from the Senate while trying to persuade developing countries to commit to limits on their emissions, which is unlikely for the foreseeable future (Pope 1998).

The controversial and seemingly self-serving positions of the United States at the COP VI were apparently calculated to make a final version of the Kyoto Protocol more palatable to hesitant senators and thereby enhance its prospects for ratification. The option of making unlimited purchases of emission credits from other countries, such as Russia and Ukraine, would make it unnecessary for the United States to adopt politically unpopular policies that would sharply constrain fossil fuel use. Provisions for receiving credit for forest and agricultural sinks might appeal to senators from timber and farm states.

Nongovernmental advocacy groups play a greater role in the U.S. policy-making process than in most countries, in part because the division of power among the executive, legislative, and judicial branches offers them more possibilities for influencing governmental decisions. Environmental groups, such as the Sierra Club, World Watch, Greenpeace, Friends of the Earth, and the Union of Concerned Scientists, have been striving for years to inform the public of the seriousness of the threat of climate change and to lobby the administrative and legislative branches of the federal government to work for strong international agreements to address the problem. The possibility that the United States would commit to mandatory emission reductions has energized numerous **nongovernmental organizations (NGOs)** whose interests might be adversely affected by policies designed to curb GHG emissions from the use of fossil fuels. A coalition of more than thirty such groups representing industries, such as coal and oil producers and automobile manufacturers, and consumer unions protecting the interests of seniors and various minorities sponsored a $13 million media blitz in 1997 to persuade Americans that the Kyoto Protocol would be unfair to the United States because developing countries would not be required to reduce their emissions.

Thus, regardless of whatever form the Kyoto Protocol might take, the chances for its ratification will be remote unless the U.S. public is convinced of the seriousness of the problem. Public opinion polls have repeatedly shown that a sizable majority of Americans are concerned about global warming and favor ratifying the Kyoto Protocol. Surveys also reveal, however, that the environment ranks rather low on the list of priorities of U.S. voters and, among environmental issues, climate change tends be of less concern to them than other more local and immediate problems, such as air and water pollution. Furthermore, Americans seem to be less persuaded than Europeans of the urgency of the threat of climate change or the need for potentially costly responses (Skolnikoff 1997).

Among the reasons for the apparent ambivalence of Americans about climate change is the exposure they have had in recent years to the views of a rather small group of scientists who have been outspoken in expressing skeptical views on the science of climate change. These scientists, some of whom have had their activities supported by industrial groups opposed to any limits on the use of fossil fuels, have received extensive coverage by U.S. media committed to presenting opposing views on public issues. They have also been given ample opportunities to testify before congressional committees by chairs who are openly hostile to the Kyoto Protocol. These skeptics appear to have had considerable success in countering the warnings of the mainstream scientific community expressed in the reports of the IPCC, and to have left the U.S. public confused about the state of science on the problem and whether immediate action to address it is warranted (Gelbspan 1997; Pearce 1997).

There also are more general features of the U.S. political culture that give politicians reason to pause before making commitments to significant GHG emission reductions. One is the strong aversion of Americans to new or increased taxes of the magnitude that would be necessary to significantly alter their energy consuming habits. Europeans are generally more tolerant of taxes, as is apparent from energy taxes in their countries that are currently several times higher than those in the United States. Americans are also much more committed than Europeans to personal freedom, such as on choosing the size and location of housing and the size of automobiles, and accordingly are more resentful of governmental regulations that may limit their choices. Achieving the reductions that would be necessary to accomplish the emission reductions mandated by the Kyoto Protocol would require lifestyle changes dictated by strong governmental policies that politicians fear would be highly unpopular among their constituencies.

Conclusion

As we enter a new century, the prospects are not very favorable for creating a strong international regime that could effectively limit global climate change. Although a promising first step, the Kyoto Protocol, as finalized in Bonn in 2001, will only slow down the rate at which GHGs are accumulating in the atmosphere, even in the unlikely event that the United States reverses course and ratifies and implements the agreement. A reduction of GHG emissions of at least 60% will be needed just to stabilize concentrations of these gases in the atmosphere, much less return them to preindustrial levels. Population growth in the developing world combined with aspirations for economic development and reduction of poverty make it highly probable that GHG emissions will continue to rise rapidly in the Third World, thus offsetting any reductions achieved in the developed world.

The basic challenge is to persuade sovereign countries to consent to making substantial sacrifices to provide a common global good in the form of a stabilized climate. The task would be easier if the countries most responsible for the problem of human-induced climate change also anticipated suffering the most serious impacts. This is not the case, however, as the United States and other major industrialized emitters of GHGs are less vulnerable to these impacts and have more resources available for adapting to them. Policy makers are especially reluctant to commit their countries to potentially costly and disruptive adjustments in the short run for benefits that may not be very apparent for decades. Ironically, the developing countries and the small island nations, which are the most vulnerable to climate change and its impacts and the least able to adapt to them, are responsible for a relatively small proportion of GHG emissions. Thus, it is unlikely that a stronger international climate change regime will be developed unless high visibility climatic and environmental events occur that can be unambiguously linked to human pollutants and have serious consequences for the United States and the other industrialized countries that have lagged in their commitment to address the problem.

REFERENCES

Acharya, A. 1995. Small islands: Awash in a sea of troubles. *World Watch* 37 (December):24–33.

Benedick, R. E. 1998. *Ozone diplomacy: New directions in safeguarding the planet,* enlarged ed. Cambridge, Mass.: Harvard University Press.

Bodansky, D. 1993. The United Nations Framework Convention on Climate Change: A commentary. *Yale Journal of International Law* 18:451–558.

Christoff, P. 1998. From global citizen to renegade state: Australia at Kyoto. *Arena Journal* 10:113–27.

Dunn, S. 1998. Can the North and South get in step? *World Watch* 40 (November/December):19–27.

Engelman, R. 1995. Imagining a stabilized atmosphere: Population and consumption interactions in greenhouse gas emissions. *Journal of Environment and Development* 4:111–40.

Gelbspan, R. 1997. *The heat is on: The high stakes battle over earth's threatened climate.* Reading, Mass.: Addison-Wesley.

Gore, A. 1992. *Earth in the balance: Ecology and the human spirit.* Boston: Houghton Mifflin.

Intergovernmental Panel on Climate Change (IPCC). 1990. *Climate change: The IPCC Scientific Assessment.* Cambridge: Cambridge University Press.

Janis, M. W. 1999. *An introduction to international law,* 3d ed. Gaithersburg, Md.: Aspen Publishers.

Jurgielewicz, L. M. 1996. *Global environmental change and international law: Prospects for progress in the legal order.* Lanham, Md.: University Press of America.

Krasner, S. D., ed. 1983. *International Regimes.* Ithaca, N.Y.: Cornell University Press.

Lanchbery, J. 1998. Expectations for the climate talks in Buenos Aires. *Environment* 18 (October):18–20,42–45.

Litfin, K. T., ed. 1998. *The greening of sovereignty in world politics.* Cambridge, Mass.: MIT Press.

Malanczuk, P. 1997. *Akehurst's modern introduction to international law,* 7th ed. London: Routledge.

Muldoon, J. P., Jr., J. F. Aviel, R. Reitano, and E. Sullivan, eds. 1999. *Multilateral diplomacy and the United Nations today.* Boulder, Colo.: Westview Press.

Ott, H. E. 1998. The Kyoto Protocol: Unfinished business. *Environment* 40 (July/August):18–20, 41–45.

Pearce, F. 1997. Greenhouse wars. *New Scientist* (July 10):38–43.

Pope, C. 1998. A year after Kyoto pact's completion, the political heat is unabated. *Congressional Quarterly* 56(46):3175–77.

Ringius, L. 1999. The European Community and climate protection: What's behind the empty rhetoric. Oslo: Center for International Climate and Environmental Research.

Rowlands, I. H. 1995. The climate change negotiations: Berlin and beyond. *Journal of Environment and Development* 4:145–64.

Skolnikoff, E. B. 1997. Same science, differing policies: The saga of global climate change. MIT Joint Program on the Science and Policy of Global Change, Report No. 22.

Soroos, M. S. 1997. *The endangered atmosphere: Preserving a global commons.* Columbia: University of South Carolina Press.

Werksman, J., ed. 1996. *Greening international institutions.* London: Earthscan.

Williams, M. 1997. Institutions for global environmental change. *Global Environmental Change* 7:295–98.

Young, O. R. 1994. *International governance: Protecting the environment in a stateless society.* Ithaca, N.Y.: Cornell University Press.

SUGGESTED READINGS

Brown, L., and J. Mitchell. 1998. Building a new economy. In *State of the World.* New York: W. W. Norton.

Gelbspan, R. 1997. *The heat is on: The high stakes battle over earth's threatened climate.* Reading, Mass.: Addison-Wesley.

Jacoby, H. D., R. G. Prinn, and R. Schmalensee. Kyoto's unfinished business. *Foreign Affairs* (July/August 1998):54.

Rowlands, I. H. 1995. The climate change negotiations: Berlin and beyond. *Journal of Environment and Development* 4:145–64.

Skolnikoff, E. 1999. The role of science in policy: The climate change debate in the United States. *Environment.* 41(5):16–25.

Soroos, M. S. 1997. *The endangered atmosphere: Preserving a global commons.* Columbia: University of South Carolina Press.

CHAPTER 7

The Long Road Ahead

CONCLUDING
THOUGHTS ON
CLIMATE CHANGE

Sharon L. Spray
and
Karen L. McGlothlin

What resonates most clearly from the essays in this volume is that climate change will likely be the most complex environmental challenge, both scientifically and socially, of the twenty-first century. We cannot hope to address the problem without an understanding of how scientific uncertainty shapes the economic and policy debate, and how global cooperation hinges on cultural and historic patterns of behavior. Although the authors in this volume have discussed these issues previously, we would like to leave you with some additional thoughts on the difficulty and urgency of linking scientific knowledge with positive environmental outcomes.

Much of what is discussed about climate change in media is a debate over the certainty of science. Major snowstorms lead to stories that suggest the planet could be cooling. Heat waves bring articles that suggest the planet is warming. It is no wonder that Americans who rely on the news media for their information on climate change would perceive that there is far more uncertainty about the issue of global warming than there is in the scientific community. The climate change story as relayed in the popular press is one that favors scientific disagreement over scientific consensus regardless of the strength of the scientific evidence or consensus in the scientific community. The search for the "scoop," or the "interesting" story, results in more attention to the flamboyant or the lone critic of mainstream scientific research on climate change than a thoughtful discussion of the steady long-term collection of data by scientists across disciplines that has accumulated for decades.

By giving equal time to global warming critics and hired industry skeptics opposing the global scientific community, the media contribute to conditions of citizen confusion and complacency in addressing this serious global environmental challenge. At the very least, highlighting controversy over consensus provides latitude for citizens to continue their lives without changing personal behavior and for business and industries to avoid the high initial costs of investment in new environmentally sound technologies regardless of the long-term benefits.

If the potential consequences of nonaction were less urgent, the media's role in the climate change debate would matter less. But the stakes are high. Climate change, if unabated, will profoundly affect our environment and its inhabitants, including (but not limited to) forest migration, desertification, the erosion of coastal areas, and an increase in the salinity of estuarine areas and coastal freshwater aquifers. Organisms that are unable to adapt to changes in their environment will be faced with the threat of extinction, leading to a continued decrease in the earth's biodiversity.

As the authors in this volume illustrate, the science of climate change is strong. We know that worldwide the emissions of carbon dioxide (CO_2) are continuing to increase at a faster rate than the concentration of CO_2 increased during the late nineteenth and early twentieth centuries (Karl et al. 2000). A recent investigation conducted by two British scientists

The Long Road Ahead   147

yielded results that indicated that present CO_2 levels are at a 20-million-year high (Pearson and Palmer 2000). We have evidence that indicates that increases in the concentration of atmospheric CO_2 have historically been accompanied by corresponding increases in average temperature (see Dobson, this volume). It should then come as no surprise that 1997 and 1998 were two of the hottest years on record and that this past decade, the 1990s was the hottest decade recorded.

Because no one can readily detect global increases in temperature on a daily basis, it is easy to ignore the potential impacts of our warming climate. There is already evidence, however, that the ranges of distribution of species of some plants and animals are expanding northward in response to average temperature increases (see Herrod-Julius and McCarty, this volume). The real challenge for addressing climate change may be that by the time most people recognize discernible changes in their surroundings, we may have reached a point where the effects are irreversible.

It is true that the scientific community does not yet fully understand climate change. New information is emerging all the time. For example, scientists recently discovered a new greenhouse gas in our atmosphere (Sturges et al. 2000). This gas, trifluoromethyl sulfur pentafluoride (SF_5CF_3), has been detected in small concentrations in our upper atmosphere and is able to trap heat far more effectively than any of the known greenhouse gases. Evidence indicates that this gas is anthropogenic in origin—possibly a by-product of the use of high-voltage equipment—and has been accumulating for approximately forty years. Although one might expect that new evidence such as this would lead to policy action, this is not necessarily the case. Opponents often point to new discoveries to strengthen their claim that our scientific understanding of climate change remains inadequate for policy action.

Whether we are experiencing global climatic change, however, is no longer at issue. Scientific data indicate that we are. What remains questionable is *how* climate change will transform our planet if we fail to take action. Possible outcomes include an increase in average surface temperature between 1.4° and 5.8°C (2.5° and 10.4°F) over the course of the next century. Increasing temperatures of this magnitude will inevitably bring a cascade of changes, including altering the geographical distributions of many organisms (see Herrod-Julius and McCarty, this volume). One likely outcome of increasing surface temperatures will be the expansion of seawater, as well as the melting of polar ice caps, leading to sea level rise and to coastal erosion, both of which have the potential to leave millions of people around the world as "environmental refugees."

There will also be increased risks for human health. One potential risk could come from a widespread increase in the instances of insect-borne diseases, including malaria, encephalitis, yellow fever, and others (Epstein 2000). Hotter temperatures will likely lead to expansions of the geographical ranges of the insects that carry these diseases, which will increase the number of people affected. Already, we have seen in the northeast United States the occurrence of the west Nile virus, a mosquito-borne virus that can cause encephalitis in humans and that had not been detected in the Western Hemisphere before 1999. By the fall of 2001, this virus was detected as far south as Georgia. Additional health concerns include increasing numbers of deaths due to heat exposure, disruptions in food production, and intense weather-related disasters. Although there is no way to be certain about what global climate change may bring, the possibilities are frightening.

The nature of science and scientific inquiry is such that not all questions, including some questions associated with global climate change, can be completely and definitively answered. Scientists cannot look into a crystal ball and reveal the future, but they can, through carefully planned experiments and observations, gather data that may be helpful in predicting the outcomes of particular activities. The political question is whether there is *enough* scientific evidence to move policy makers and citizens to make choices that will save us from the irreversible consequences of our own behavior.

Fortunately, advances in technology are making it easier for citizens to make environmentally sound decisions without dramatic changes in lifestyle. Better insulation of homes, the use of new energy efficient compact fluorescent lightbulbs, and energy-efficient appliances can save hundreds of kilowatt hours of energy and reduce the amount of CO_2 released into the atmosphere. Innovative use of hydrogen fuel cells, such as those used to power the Toyota Prius and the Honda Insight, shows promise that even the ever-popular sport utility vehicle can be designed to operate efficiently.

Many businesses and industries are realizing the potential savings acquired through the use of environmentally sound technology and innovative sources of power generation. For example, "[i]n Germany and Switzerland, new office buildings are incorporating photovoltaic cells into the windows of south-facing facades," allowing individual buildings to generate their own energy resources (Brown and Mitchell 1998). In the United States industries are reaping the benefits of cogenerated power, the process by which the waste heat from electricity generation is made available for industrial processes by installing small boilers and electric generators within plants (Smith 2000). And countries around the world, including the United States, are just beginning to see the potential benefits of advancements in renewable energy technologies such as wind, solar, and geothermal power.

The real issue for most Americans is not whether we should continue using fossil fuels. There is nothing essentially unifying about an economy based on the use of fossil fuels. The attachment to a fossil fuel-based economy is about lifestyle. Environmental quality continues to rank high in public opinion polls that ask Americans about issues of national importance. There is no evidence to suggest that Americans will not favor more environmentally sound choices if those choices meet the needs of their lifestyle and are dependable, economical, and readily available. The availability of relatively cheap, fossil fuel-based energy and a growing economy have enabled most citizens to buy larger homes, more powerful vehicles, and more electronics and home appliances. Suggesting that people will want to give up these things is naive. We should be talking about policies that increase research and development of environmentally sound technologies and identify strategies to implement and bring down the costs of these emerging technologies.

On the whole, the majority of Americans still do not fear the outcome of climatic change enough to modify their own behaviors or to pressure policy makers to make hard decisions about economic and regulatory policies that would decrease our dependence on fossil fuels. A gasoline price hike of fifty cents a gallon in the summer of 2000 resulted in angry emails and phone calls to Congress demanding reductions in fuel costs, while at the same time Americans were buying more sport utility vehicles than any other type of vehicle on the market. Europeans literally took to the streets during the fall of 2000 to protest gas prices in several European countries. Such visible and angry objections to increasing gas prices will inevitably dampen efforts to attach additional carbon taxes or pump taxes designed to decreased consumption, especially if consumers fail to connect their own consumption with environmental damage.

Creating a sense of urgency is difficult without a focusing event to galvanize public concern. The global warming "story" lacks the equivalent of a discernible hole in the ozone layer or the deaths of hundreds people in a single day from a chemical accident to mobilize government, industry, and citizen action. What we have with climate change are subtle changes to our global landscape outside the vision of most people's everyday lives. It is a thawing of the permafrost at the North Pole, the cracking of glaciers at the South Pole, longer summers and shorter winters, the disappearance of species in one area and the appearance of insects in another. Climate change has resulted in a collective group of small disturbances that tell a larger story of environmental harm. In this sense, the overt consequences of climatic changes are like the science that underpins our understanding of the problem.

Our understanding of climate change evolved slowly through the collection of scientific data from the atmosphere, the earth, and oceans. It has been a process of replication, modification, and reinterpretation. In science this is not the basis of certainty because scientists understand that their reliance upon evidence does not yield absolute truths. Science is a dynamic and controversial process in which explanations are offered on the basis of observations by individuals. There is room for bias and misinterpretation in the scientific process. The safeguard, however, is the fact that numerous individuals typically investigate a particular topic and their results are customarily subjected to a rigorous process of peer review. Models that challenge conventional scientific wisdom are retested, debunked, revised, reformulated, and considered within the entire body of scientific knowledge. In this sense, inaccurate or controversial studies move science forward. But whenever there is controversy—either about current conditions or future outcomes—public understanding of a topic can be clouded, creating doubt and galvanizing opposition to change. This is to a great extent what has happened in the United States, where powerful stakeholders benefiting from the current fossil fuel–based economy have shaped the political debate about climate change to one about risk of economic disruption rather than risks to our environmental health.

The United States is the world's largest consumer of fossil fuels and largest producer of greenhouse gases, but the United States has not taken a leadership role in the international negotiation process to secure global cooperation on this critical environmental challenge. Since baseline targets for CO_2 emission reductions were set in Rio in 1992, the CO_2 emissions of the United States have increased 11.7% over the 1990 emissions, which were used to establish those targets (1999 EPA data). This fact is important because the United States is the one state actor whose leadership is the most important in the development of international regimes to address large-scale environmental problems (Porter and Brown 1996). The stand taken by the United States since the election of George W. Bush as president and Dick Cheney as vice president—both men who have strong ties to oil interests—indicates that there will be little change in U.S. policy in the near future.

We also want to note that although U.S. leadership is a necessary condition for an effective environmental regime to address climate change, it is certainly not sufficient. As the authors in this book suggest, addressing climate change on a global scale is complicated by issues of national sovereignty and vast differences in culture, wealth, and industrial development as well as issues of fairness and equity. During international negotiations, developing countries have repeatedly pointed out that industrialized countries grew rich through the use of fossil fuels but are now asking newly industrializing countries to make different choices. Many involved in the policy debate believe that making such demands is hypocritical unless there is both accountability for the damage incurred as a result of the past practices of industrialized nations, and assistance given to developing countries so that they may develop with cleaner, more energy-efficient technology that only the richest countries can presently afford.

The strength of international cooperative efforts will hinge on the willingness of industrialized countries to help developing nations address issues of poverty, population, and natural-resource depletion in ways that help them develop their economies without transferring the practices that currently contribute to a warming of the planet. For these reasons, the U.S. diplomatic influence, financial resources, and technological potential will be essential in addressing climate change on a global scale.

As pointed out by the social scientists in this book, international cooperation can be achieved. We have international environmental regimes in place today that have successfully addressed ozone depletion and other pollution problems. But the global ramifications of climate change dwarf many of our previous battles. Literally billions of people and most nations of the world are contributors to the anthropogenic accumulations of greenhouse gases in the atmosphere. We are, indeed, talking about more than minimal changes in most countries' energy consumption and development patterns and more than minimal disruption to the world economy.

Reductions in greenhouse gases will require changes in agricultural patterns and the construction of more environmentally friendly homes and buildings. It will require the improved energy efficiency of vehicles and the planning of mass transportation to meet the demands of urban growth. Industries will need to reduce waste streams and economize on carbon intensive materials. It will also mean restructuring the energy sectors of most economies to favor renewable energy sources—natural gas in some cases, and in others, the use of micropower generation that can meet the needs of specific industries and isolated communities.

Cooperation between countries and throughout industry sectors will be critical in the development of climate change mitigation strategies.

Cooperative ventures proposed under the Kyoto Protocol's Activities Implemented Jointly (AIJ) program are indicators that cooperative ventures are, indeed, feasible. Current cooperative ventures include a variety of initiatives between industrialized and developing nations ranging from the capture and reuse of methane gas from landfills to the building of wind farms and biomass-powered utilities for regional energy production.

Other proposals include projects that utilize relatively simple, existing technology, highlighting the vast possibilities for change when states cooperate on even small-scale projects. A project between Mexico and Norway to provide incentives to replace approximately 200,000 ordinary, incandescent lightbulbs with compact fluorescent lightbulbs in the Mexican cities of Monterrey and Guadalajara is one such project. Because compact fluorescent lightbulbs require 25% of the energy of ordinary bulbs and produce similar or better-quality lighting, this project will significantly reduce regional electricity demand and fossil fuel emissions.

A joint venture proposed by the Netherlands and Costa Rica will potentially reduce methane emissions and water pollution substantially through the installation of water treatment facilities at coffee mills. Projects like this have spin-off effects. In this case, a project to reduce methane emissions also will facilitate improvements in sanitary conditions and reduce insects and odor from unregulated waste disposal into water resources. There will be other side benefits as well, such as increased availability of water for downstream agricultural purposes. The greatest benefit may be that this type of small, cooperative program promotes significant reductions in greenhouse gases while preserving local culture and protecting local ecosystems.

Forestry options, although highly controversial at the sixth Conference of the Parties (COP VI), also offer potentially high benefits for reducing carbon dioxide emissions with modest costs to natural ecosystems and native cultures. Forestry options currently being discussed and tested include cooperative ventures to halt or slow deforestation, programs to increase reforestation and silvicultural productivity, and the expansion of forest preserves and ecotourism. These forestry projects are not without risk, however, and must be carefully planned and implemented. Initial pilot projects indicate that without the involvement of local stakeholders in the design and management of these activities, reduced access to land, fuel, fiber, and timber simply forces deforestation and environmentally unsound practices to different locations (Intergovernmental Panel on Climate Change [IPCC] 2000).

The list of possible cooperative ventures is long. The real question is whether we can move from relatively small-scale, successful projects to large-scale, global programs to transfer technology and assist developing countries in meeting their energy needs without compromising local cultures and accumulating debt that could jeopardize other programs desperately needed to address poverty, education, and food production.

An obvious part of this equation is finding ways to finance the transfer of new technology between firms, between communities, and between countries. The tradable emissions and joint implementation programs, currently on the table (see Soroos, this volume) in formal international negotiations, would potentially facilitate some activities, but it is highly unlikely that these programs will be sufficient.

"[I]nnovative financial mechanisms involving public/private partnerships and specialized credit facilities" are also needed to finance research, development, and the transfer of technology at a scale necessary to reverse accumulations of greenhouse gases in the atmosphere (IPCC 2000). With the transfer of this technology there must also be a willingness to invest in the "human capacity" of nations. Obviously it will require the development of more skilled labor forces. This will only occur if adequate investment is made in education and training programs in countries that currently cannot afford them. It will also require that many countries develop the business, legal, management, and regulatory skills that will be necessary to ensure compliance and proper implementation of technology and related programs. Otherwise, many promising ventures will fail and ultimately discourage badly needed foreign investment and continued cooperation (IPCC 2000).

The transfer of technology will also require the building of "organizational capacities" in many developing countries. By this we mean the development of firms to manage (business and environmental) accounting, enforce laws, implement product ratings and standards, monitor trade, and

The Long Road Ahead

develop communication and information networks (including the Internet) that will facilitate the sharing of information (IPCC 2000). It will also be necessary for some countries (developed and developing) to create favorable economic and political conditions for private investment by reducing subsidies, increasing the protection of intellectual property rights, promoting competition, reducing corruption, and making sure that local citizens are involved in decisions that directly affect their lives and their local economies (IPCC 2000).

While all of these efforts will increase our ability to address global climate change, they will also increase policy success in other areas. The transfer of technology and adequate investments strategies will also lead to greater economic and environmental stability throughout the industrializing world and ultimately to the sustainability of global resources and the overall environmental health of the planet.

Addressing climate change will be difficult. The authors in this volume provide some perspective on the complexity of this environmental challenge. There are a number of uncertainties associated with the eventual impacts of climate change and our ability to restructure global energy use patterns. But waiting until we have all of the answers is not an option if we want to save our planet for future generations. We may not know the outcome of the present rate of climatic change, but we do know that we will pay greatly for our inaction. For this reason, the continued research of scientists, social scientists, and humanists is important. Equally important is finding ways to link these perspectives to initiatives that facilitate efforts to address this global environmental challenge while we still have the ability to do so.

REFERENCES

Intergovernmental Panel on Climate Change (IPCC). 2000. Methodological and technical issues in technology transfer: Summary for policymakers. A special report of Working Group III of the Intergovernmental Panel on Climate Change. Available online (visited September 2000) at: http://www.ipcc.ch.

————. IPCC Second Assessment Report: Climate Change 1995. Summary for policymakers: The economic and social dimensions of climate change. IPCC Working Group III. Available online (visited September 2000) at: http://www.ipcc.ch.

————. Land use, land-use change, and forestry. Summary for policymakers. A special report of the Intergovernmental Panel on Climate Change. Available online (visited September 2000) at: http://www.ipcc.ch.

Brown, L., C. Flavin, H. French, et al. 1998. *State of the world.* New York: W. W. Norton.

Brown, L., and J. Mitchell. 1998. Building a new economy. In *State of the world.* New York: W. W. Norton.

Epstein, P. R. 2000. Is global warming harmful to health? *Scientific American* 283:50–57.

Karl, T. R., R. W. Knight, and B. Baker. 2000. The record-breaking global temperatures of 1997 and 1998: Evidence for an increase in the rate of global warming? *Geophysical Research Letters* 27(5):719

Pearson, P. N., and M. R. Palmer. 2000. Atmospheric carbon dioxide concentrations over the past 60 million years. *Nature* 406:695–99.

Porter, G., and J. W. Brown. 1996. *Global environmental politics.* Boulder, Colo.: Westview Press.

Smith, Z. A. 2000. *The environmental policy paradox,* 3d ed. Upper Saddle River, N.J.: Prentice-Hall.

Sturges, W. T., T. J. Wallington, M. D. Hurley, K. P. Shine, K. Sihra, A. Engel, D. E. Oram, S. A. Penkett, R. Mulvaney, and C. A. M. Brenninkmeijer. 2000. A potent greenhouse gas identified in the atmosphere: SF_5CF_3. *Science* 289:611–13.

U.S. Environmental Protection Agency (EPA). 1999. Recent trends in U.S. greenhouse gas emissions. *The EPA global warming site,* http://www.epa.gov/globalwarming/emissions/trends.html.

Glossary

accrete - to cause to adhere.

advection - the horizontal, or lateral, transport of a substance.

albedo - reflectivity of a surface; materials with high albedo (such as ice and snow) reflect more incoming radiation back to space, while materials with low albedo (soil, rock) absorb more incoming radiation and heat up.

Annex I countries - industrialized countries listed in Annex I of the Framework Convention on Climate Change. Annex I countries agreed to binding emissions reductions and to provide detailed reports of plans to reduce emissions and enhance carbon sinks.

anthropocentric - a human-centered evaluation of the environment solely in terms of how it can fulfill human needs and desires.

anthropology - the study of humankind around the globe and through time.

archaeology - a subdiscipline of anthropology that studies prehistoric and historic people through their material culture.

behavioral and evolutionary ecology - a subfield of anthropology that considers how humans evolve behaviorally or physically to a given environment and in relation to changes in that environment over time.

benefit-cost analysis - a tool used by economists to evaluate policy options in terms of their measurable, monetary costs and benefits.

biocentric - a philosophical view of the environment that examines the human-environmental relationship in terms of the intrinsic value of nature rather than the environment's utility to humans.

biome - a group of ecosystems that have a similar vegetation type and similar climatic conditions.

biosphere - the overall ecosystem of the earth, including the sum of all biomes.

blackbody radiator - an object that absorbs all forms of radiation striking its surface and acts as a perfect emitter of radiation, radiating energy at particular wavelengths dependent upon temperature. At a given temperature, the body emits a large range of frequencies

157

158   GLOBAL CLIMATE CHANGE

with the maximum wavelength shifting as the temperature increases. The emission is referred to as the blackbody spectrum. Blackbody radiators behave uniformly; all those with the same temperature emit the same spectrum.

buffering - circumstances that counteract or resist an initial change.

capital flight - the process of business firms leaving a country in order to find lower wages or fewer regulatory restrictions.

carbon dioxide (CO_2) - a chemical compound produced by oxidation of carbon, such as occurs during respiration or burning of organic materials; carbon dioxide is a greenhouse gas that absorbs and retains heat radiating from the earth's surface.

chlorofluorocarbons (CFCs) - any of several organic compounds composed of carbon, fluorine, chlorine, and hydrogen. CFCs are manufactured under the trade name Freon (q.v.). Developed during the 1930s, CFCs found wide application after World War II. These halogenated hydrocarbons, notably trichlorofluoromethane (CFC-11, or F-11) and dichlorodifluoromethane (CFC-12, or F-12), have been used extensively as aerosol-spray propellants, refrigerants, solvents, and foam-blowing agents. Because of a growing concern over stratospheric ozone depletion and its attendant dangers, a ban was imposed on the use of CFCs in aerosol-spray dispensers in the late 1970s by the United States, Canada, and the Scandinavian countries. In 1990, ninety-three nations agreed to end production of ozone-depleting chemicals by the end of the century; in 1992, most of those same countries agreed to end their production of CFCs by 1996.

circumpolar vortex - a circular wind over the poles.

clean development mechanism (CDM) - the component of the Kyoto Protocol that would allow firms in wealthy countries to claim greenhouse gas emission reduction credits for transferring clean technology to developing countries.

climate proxy - a data set that provides a historic (or prehistoric) record of climate, such as tree rings, oxygen isotopes, coral growth bands, and thicknesses of annual ice layers.

command-and-control (CAC) regulation - a type of environmental regulation that requires that firms install only particular types of government-approved cleanup technologies to achieve emission reductions.

condensation nuclei - small, suspended, solid or liquid particles in the atmosphere on which water vapor may condense to begin cloud formation.

conservation biology - the scientific study of the factors that affect the maintenance, loss, and restoration of the plant and animal species that contribute to biological diversity.

convention - a treaty negotiation open to all states.

Glossary

159

cryosphere - that portion of the climate system consisting of the world's ice masses and snow deposits. This includes the ice sheets, ice shelves, ice caps and glaciers, sea ice, seasonal snow cover, lake and river ice, and seasonally frozen ground and permafrost.

cultural ecology - a subfield of anthropology that considers how the culture of a human group is adapted to and shaped by the natural resources of an environment and to the presence of other human groups.

cultural models - ways in which different cultures configure their values and beliefs; used here specifically to refer to the ways different cultures perceive the environment.

culture - learned behaviors and beliefs that we think and do that are not instinctual.

discount rate - the rate at which future costs and benefits of a project are discounted. High discount rates reduce the value of future costs and benefits more than low discount rates.

discounting - in benefit-cost analysis, the process of valuing the future benefits and costs of a proposed policy less than the present benefits and costs. The rationale for discounting is that spending dollars on a project today has an opportunity cost in the form of foregone, productive investment in other areas.

ecological anthropology - draws on all subfields of anthropology and other disciplines to consider past and present relationships between societies and their environments as mediated through culture.

ecological community - the species of plants and animals that occur together in a place.

ecological model - a formal mathematical, verbal, or graphical representation of how ecological variables interact.

ecology - the scientific study of the relationship of organisms to their physical and biological environment that seeks to understand the factors that determine the distribution and abundance of plants and animals.

ecosystem - a community of plant and animal species and the nonliving environment that they inhabit.

ecosystem function - the interactions of the parts of an ecosystem, includes the exchange of water, nutrients, and energy among the living and nonliving parts of an ecosystem.

ecosystem services - the conditions and processes through which natural ecosystems, and the species that make them up, sustain and fulfill human life.

ecosystem structure - the physical form of the ecosystem; the living and nonliving parts and how they fit together.

enculturation - the process by which individuals learn or acquire their society's cultural models.

Enlightenment - a philosophical movement of the eighteenth century that stressed human reason and scientific study as the basis of discovery. The period is marked by significant changes in politics, religion, and educational practices.

environmental toxicology - the branch of science that studies the fate of potentially toxic substances in the natural environment and how those substances impact natural ecosystems.

ethnoecology - a subfield of anthropology that describes people's conceptual models of their environment—such as how they classify and interact with plants, animals, landforms, water bodies—which subsequently effects the way in which different cultures attribute significance and divine attributes to environments.

ethnography - a strategy used by anthropologists to describe a culture through simultaneous participation and observation within a community supplemented by interviews and the collection of oral histories.

ethnohistory - the anthropological study of past societies through the study of written sources, oral histories, artwork, and photography.

externality (negative) - a cost of a transaction not borne by the seller or the buyer.

faculae - intense electromagnetic regions of high-energy output that surround a sunspot.

feedback - when a system responds to a change in such a way that the change is amplified (positive feedback) or damped (negative feedback); important to understanding and estimating how global climate will respond to changes.

flexibility mechanisms - options provided in the Kyoto Protocol that permit Annex I states to achieve commitments to limit greenhouse gas emissions by means other than directly reducing their own emissions, such as emission trading, joint implementation, group bubbles, and the clean development mechanism.

flux model - a mathematical analysis of the transport of materials from one state to another or from one place to another. Often used to study the flow of geologic or atmospheric compounds, including water and carbon dioxide, around the earth.

forcing - a system in equilibrium that is perturbed by an external force. Positive feedback in climate systems is an example of a forcing function.

fossil fuels - energy sources composed of the altered remains of long-dead plants and animals; examples are coal, oil, natural gas, and petroleum, all of which produce carbon dioxide when burned.

framework convention - an initial international agreement that acknowledges a problem but typically contains few specific actions by the parties to mitigate the problem. Such

Glossary

agreements establish the foundation for negotiating later agreements that commonly take the form of protocols.

general circulation model (GCM) - a computer model of global climate that divides the surface of the earth into a grid and estimates a variety of climate parameters for and interactions between grid cells; useful in predicting future climate change.

global warming - a rise in average global temperature; currently, a predicted increase of several degrees over the next century due to increased levels of greenhouse gases in the atmosphere.

greenhouse effect - a process that warms the surface of the earth; visible light and other radiation from the sun can travel freely through the earth's atmosphere and heats up the earth, but heat from the earth is partially trapped by greenhouse gases in the atmosphere, so the earth retains some of its escaping heat.

greenhouse gases (GHGs) - any of a number of gases (e.g., methane and carbon dioxide) that absorb and retain heat radiated from the earth's surface, producing the greenhouse effect.

Hadley cell - low-latitude air movement toward the equator that with heating, rises vertically, with pole-ward movement in the upper atmosphere. This forms a convection cell that dominates tropical and subtropical climates.

heterarchy - the relation of elements to one another when they are unranked or when they possess the potential for being ranked in a number of different ways (Crumly 1995).

historical ecology - an anthropological approach to studying human-environmental relationships over time; expressly focusing on how past human cultures responded to and impacted their natural environments.

Holocene - our current geological period.

ice core - a cylinder of ice obtained by drilling a hole in an ice cap, usually in a mountain glacier or in the ice caps of places such as Greenland or Antarctica; the ice normally contains annual or seasonal layering and bubbles of gas from the atmosphere at the time the ice formed.

Industrial Revolution - radical social and economic changes that occurred in the late eighteenth century as a result of the mechanization of production systems that shifted small-scale, home manufacturing to large-scale, factory production.

incentive-based (IB) regulation - a type of environmental regulation that first puts a price tag on each unit of pollution produced by a firm and then provides firms with flexibility in choosing how to reduce emissions. IB regulation includes both marketable permits systems and pollution taxes.

162 GLOBAL CLIMATE CHANGE

infrared (IR) spectrum - region of light emissions with wavelengths of 700 nm and greater.

international customary law - unwritten norms of behavior observed by governments in their relations with one another that are considered to be legal obligations. These norms are binding on all states except those that have explicitly rejected them as they evolved.

international regime - the international governance, including international institutions, norms of customary law, treaties, and resolutions, that has been established to address a certain problem or issue, such as ozone depletion and climate change.

Intertropical Convergence Zone (ITCZ) - the equatorial region where the Northeast Trade Winds (Northern Hemisphere) and Southeast Trade Winds (Southern Hemisphere) flow together and are characterized by strong upward motion and heavy rainfall. The ITCZ is most clearly defined over the eastern portions of the Pacific and Atlantic oceans.

isotopic fractionation - in the study of stable isotopes, the tendency of different atoms of the same element to react differently based on their atomic weight. For example, water molecules containing oxygen atoms with ten neutrons (oxygen 18) will not evaporate as readily as water molecules containing oxygen atoms with eight neutrons (oxygen 16), which means water vapor will tend to be enriched in oxygen 16 relative to oxygen 18.

isotopic ratio - a ratio of atoms/molecules with the same number of protons, but a different number of neutrons in their nuclei.

joint implementation - a flexibility mechanism of the Kyoto Protocol that permits developing (Annex I) countries to comply with commitments to limit their greenhouse gas emissions by investing in projects that reduce emissions in other developed countries presumably at lower cost.

Judeo-Christian - religious traditions that interprets human-environmental relationships as divinely ordained.

Kyoto Protocol - an international treaty signed in Kyoto, Japan in 1997 that would commit the developed countries to reduce emissions of carbon dioxide and other greenhouse gases to a total of 5% below 1990 levels, averaged over the period 2008–2010. The treaty has not yet been ratified by most countries, including the United States.

latent heat flux - heat carried into the atmosphere by water vapor that has evaporated at the surface.

latent heat of condensation/fusion - heat absorbed or given off by condensation or fusion without undergoing a change in temperature.

lithosphere - the part of the earth lying below the crust (30 km deep) that contains special hydrated minerals (rocky minerals containing water within their crystal structure) that

Glossary

have the ability to slide against each other and enable movement of the crust and tectonic plates.

longwave (LW) radiation - electromagnetic radiation longer than about 5 microns and considered mid- and far-infrared radiation or heat.

marketable permit system (cap-and-trade system, tradable permit system) - a system of environmental regulation in which firms either are issued or purchase a set number of emission permits and are given flexibility in determining how they will achieve their permitted pollution level. Firms that achieve pollution reduction below their permitted level can sell the excess permits. Firms that fail to make their permitted pollution targets can then purchase the extra permits.

Maunder minima - Time of very low (to zero) sunspot activity between 1645 and 1715 A.D.

Medieval Climatic Optimum - the period between 900 A.D. and 1200 A.D. in which climate facilitated the northwest expansion of the Norse Vikings into Greenland and beyond.

Mesolithic - the middle stone age. A transitional period between the old stone age (paleolithic) and the beginning of agriculture in the new stone age (neolithic), ranging roughly from 15,000 to 8,000 years ago during a time of major climatic changes around the world.

methane (CH_4) - a chemical compound produced (among other ways) by respiration of anaerobic bacteria in places such as swamps, rice paddies, and the intestines of ruminant animals; methane is a greenhouse gas that absorbs and retains heat radiating from the earth's surface.

methane gas hydrate - frozen methane retained in deep-sea sediments; vast deposits of gas hydrate exist and might add large amounts of methane to the atmosphere if the earth warms and they melt.

Montreal Protocol - an international treaty signed in 1986 that laid the groundwork for the successful halt in production of chlorofluorocarbons (CFCs), which destroy the stratospheric ozone layer.

near infrared (NIR) radiation - light of wavelengths above 800 nm (red) to about 2000 nm. Not visible to the human eye, but can be felt as heat.

niche - the set of all environmental conditions that define where a species can live and reproduce.

nongovernmental organizations (NGOs) - organizations whose members are normally not governments, but private individuals or groups, such as advocacy groups, scientific unions, and trade associations.

164   GLOBAL CLIMATE CHANGE

ozone (O_3) - a reactive chemical compound consisting of three oxygen atoms; ozone high
in the atmosphere absorbs incoming ultraviolet radiation from the sun, shielding the
earth's surface from this harmful radiation.

ozone depletion - reduction of the concentration of ozone in the upper atmosphere, most
markedly in polar regions; tied to human use of chlorofluorocarbons and other chemicals that react with and destroy ozone.

ozone hole - rapid thinning of the polar stratospheric ozone layer during the springtime;
first observed over Antarctica where strong circumpolar currents confine the ozone
destruction to a well-defined region.

paleoclimatology - the study of ancient climates through climate proxies, including ice
cores, tree rings, and coral growth data.

paleoecology - the branch of ecology concerned with identifying and interpreting the
relationships of ancient plants and animals with their environment.

phenology - the timing of periodic biological events, often in relation to the seasons.

photosynthesis - the process by which plants use the energy of the sun to combine CO_2
and water into sugars, producing oxygen as a by-product.

phytoplankton - photosynthetic, microscopic algae that are suspended in that part of the
water column in which light is able to penetrate.

polar stratospheric clouds (**PSCs**) - optically thin clouds that appear regularly in the winter polar stratosphere at 12–27-km altitude.

pollution tax (**effluent fee, emission charge**) - a system of environmental regulation in
which firms are charged a fee for each unit of pollution that they emit.

precautionary principle - a normative rule for guiding development decisions for unique
resources, and/or when uncertainty renders benefit-cost analysis useless. The principle
states: To preserve the welfare of future generations, unique forms of natural capital
should be protected unless the costs of doing so are prohibitively high.

protocol - an international treaty that supplements an earlier framework treaty or convention and usually requires more specific actions by the parties to ameliorate a problem.

public good - a good from whose enjoyment people cannot be easily excluded. (*Note*:
Public goods need not be common property, for example, a radio station.)

ratification - a state's official acceptance of an international agreement that conveys its
willingness to be bound by its provisions. A treaty "comes into force" when a specified
number of states have submitted their ratifications.

reservoir - in a flux model, a state or place in which the material being studied resides for
a time (e.g., the ocean for water, or the atmosphere for carbon dioxide).

Glossary

residence time - in a flux model, the average amount of time a quantity of the material being studied will stay in a particular reservoir.

resolution - a document adopted by majority vote in an international body such as the General Assembly, which may also be referred to as declarations, statements of principles, and action plans. These documents are generally not binding on member states, even those that voted for them.

Roman Climatic Optimum - the period of 300 B.C. to 300 A.D. in which climate facilitated the northward movement of the Roman Empire.

sensible heat flux - heat transfer to the overlying atmosphere due to solar heating of the earth's surface minus the latent heat flux (*see* latent heat flux).

shortwave (SW) radiation - electromagnetic radiation from the near ultraviolet (UV-A, 320 nm) to the near infrared (<5 microns).

sink - in a flux model, a way to transport the material being studied out of a reservoir (e.g., evaporation of water out of the ocean).

soft law - documents adopted in international bodies, such as resolutions and declarations, whose provisions are generally not considered to be binding on states. Such documents may set the stage for negotiating treaties and, if widely accepted, may be regarded as international law.

solar constant - the value of the total power received by the earth from the sun.

solar maximum - a year of maximal sunspot activity.

solar minimum - the lowest level of annual sunspot activity.

source - in a flux model, a way to transport the material being studied into a reservoir (e.g., rain falling into the ocean).

sovereign - not subject to higher authority. Nation-states are sovereign in the sense that they are not subject to higher political authority that is empowered to impose rules on them without their consent.

species abundance - the number of individuals of a species in a specified area or population.

species distribution - the geographic area where members of species live and reproduce.

stable isotopes - atoms of a chemical element (e.g., carbon or oxygen) that have differing numbers of neutrons but are not radioactive; the ratios of various stable isotopes of carbon (^{12}C and ^{13}C) and oxygen (^{16}O and ^{18}O) and other elements in a variety of materials are used as climate proxies.

steady state - in a flux model, when the movement of material into a reservoir is balanced by the movement of that material out of the reservoir (i.e., when sources equal sinks). A reservoir in steady state will not change size over time.

stratosphere - the region of the atmosphere where temperature drops uniformly with increasing altitude, a very stable region.

stratospheric ozone (O_3) - ozone that naturally forms in the stratosphere and has a peak concentration around 20 km altitude.

sustainability - a state of the world in which humans have reduced both pollution and depletion of resources so that, on balance, living standards for the typical (median) member of any future generation will not decline below their current level.

treaty - a formal written agreement negotiated two or more states, which may also be referred to as a convention, protocol, or charter. Such agreements spell out certain obligations and privileges of the states that ratify them and thus become parties.

tropopause - the transition from the troposphere to the stratosphere at an altitude of about 18 km.

troposphere - the region of the atmosphere from the surface to approximately 18 km upward.

ultraviolet radiation - light of wavelength below 400 nm (violet). Not visible to the human eye, but is highly energetic and causes damage to life-forms.

utilitarianism - the nineteenth-century philosophy underlying much of economic analysis. Public policy should be directed toward providing, as philosopher Jeremy Bentham put it, the "greatest good for the greatest number" of people.

visible region - light visible to the human eye: between 400 nm and 700 nm.

Vostok - a research area in Antarctica near the magnetic South Pole where many ice cores have been collected; records of layer thickness, composition of trapped air bubbles, and other parameters of the ice have provided very good climate proxies for the study of ice ages and modern climate change.

World Trade Organization (WTO) - the organization in which international trade treaties are both negotiated and adjudicated.

REFERENCES

Crumley, C. 1995. Heterarchy and the analysis of complex societies. In *Heterarchy and the Analysis of Complex Societies,* edited by R. Ehrenreich, C. Crumley, and J. Levy. Arlington, Va.: The Archeological Papers of the American Anthropological Association.

Index

abundance, 70–71
accretes, 35
acid rain, 89, 111
Activities Implemented Jointly (AIJ), 153
advection, 35
advocacy groups, nongovernmental, 139–40
"ahu" (stone heads), 92
albedo, 9–11, 12, 38, 39
Alliance of Small Island States (AOSIS), 132–33
alternative technologies, 114
amphibians, 70–71
Anasazi cliff-dwelling culture, 4
Annex I countries, 128, 129, 130, 134
Antarctica, 18–19, 34, 51, 52, 53, 67, 70
anthropocentric, 84, 87, 88, 89, 102
anthropological ecology, 84
anthropology, 82
archaeology, 82
Arctic, 53
Asia, 28
atmospheric CO_2, 18–19, 20 fig.6, 21 fig.7, 27, 33–34,
Aurora Australis, 42
Aurora Borealis, 42
Australia, 75, 96, 129, 134
Austria, 95

behavioral ecology, 83
benefit-cost (BC) economics, 105–7

Bierstadt, Albert, 88
biocentric, 84, 88, 89
biodiversity, 76, 89
biomes, 75
biosphere, 31
bird(s)
    breeding, 68–69
    cassowary, 89
    crossbills, 63, 65
    European pied flycatchers, 68
    extermination of flightless, 89
    great tit, 69
    migration, 68–69
    Pacific Ocean seabirds, 71
    penguins, 70
    ranges, 70
blackbody radiator, 38
Botswana, 96
bubble provision, 130
bubonic plague, 91–92
buffering, 10
Bush (George H.W.) administration, 138
Bush, George W., 131, 138, 151
butterflies, 69–70

$CaCO_3$. *See* calcium carbonate
calcite. *See* calcium carbonate
calcium carbonate, 17
California, 70, 71, 112
Caloianul, 92
Canada, 28, 81, 126, 129, 134
cap-and-trade system, 110, 111, 114

capital flight, 117
carbon dioxide (CO$_2$)
   atmospheric, 18–19, 20 fig.6, 21
      fig.7, 27, 33–34
   automobiles and, 50
   buffering and, 10–11
   carbon flux models and, 23–24
   double, 20 fig.6, 25
   ecological research and, 63, 64,
      71–72, 73
   economics and, 106, 107, 110
   effects of, 25–26
   fossil fuels and, 8–9
   global warming and, 20–22, 55
   greenhouse effect and, 40
   as greenhouse gas, 7
   increases in, 60, 146–47
   politics and, 133, 135–36
   reducing, 153
   role of, 48–50
   steady state and, 14
   technological advances and, 149
   United States and, 151
carbon flux models, 22–25
cassowary bird, 89
caterpillars, 69
Catlin, George, 88
Celtic culture, 94
certified emission reductions (CERs),
   130
chambers, controlled climate, 73
Changing Atmosphere Conference
   (Toronto, 1988), 126, 127
chaos theory, 11
Cheney, Dick, 151
China, 114, 116, 135, 136
Chinese dynasties, 4
chinstrap penguins, 70
chlorine reservoir compounds, 53
chlorofluorocarbons (CFCs)
   atmospheric life of, 55
   as greenhouse gas, 7
   IR radiation and, 48

Montreal Protocol and, 114, 133
   ozone and, 35, 52, 53–54
   politics and, 122, 133
Church, Frederic, 88
circumpolar vortex, 52
clean development mechanism
   (CDM), 111, 130
clean technology, 112–13
climate proxies, 15–17, 33
climatologists, 4
Clinton, Bill, 127, 138
Clinton administration, 133, 138, 139
cloud-condensing nuclei (CCN), 46
clouds, 11, 35, 39, 46, 47, 48
CO$_2$. See carbon dioxide (CO$_2$)
cogenerated power, 149
Colorado, 71, 108–10
command-and-control (CAC)
   regulation, 110, 111
condensation nuclei, 46
Conference of the Parties (COP I),
   127, 128
Conference of the Parties (COP II),
   127
Conference of the Parties (COP III),
   127, 128, 132, 134
Conference of the Parties (COP IV),
   127, 131, 139
Conference of the Parties (COP V),
   127, 131
Conference of the Parties (COP VI),
   131, 134, 139, 153
conservation, 90
conservation biology, 59
conventions, 124
coral bleaching, 71
coral reefs, 71
corporate average fuel economy
   (CAFE) standards, 112
Costa Rica, 70, 153
Cretaceous period, 4
crossbills, 63, 65
Crumley, Carole, 93, 94

cryosphere, 31
cultural challenges, 81–99
cultural ecology, 83
cultural models, 84
culture, 82

deforestation, 23–24
Denmark, 96, 130
discounting, 106
Dobson Unit (DU), 32 fig.1, 54 fig.11
Dogon, 89–90

E.I. du Pont de Nemours and
    Company, 52
Earth Radiation Budget Experiment
    (ERBE), 47–48
earth science, 4
Earth Summit (Rio de Janeiro, 1992),
    126, 127, 132
Easter Island, 92
ecological anthropologists, 82
ecological anthropology, 82–83, 84
ecological communities, 59, 71–72
ecological models, in research, 65–66
ecology, 59, 60, 61, 62–66, 82
economics, 101–18
    benefit-cost approach to, 105–7
    clean technology and, 112–13
    of global environmental issues,
        113–17
    market systems and, 103–5
    precautionary principle of, 107–8
    of regulation, 108–13
    utilitarian framework and, 102–3
ecosystem function, 60, 62
ecosystem services, 60–61, 76, 105
ecosystem structure, 60, 61, 62, 70
ecosystems, 59, 60, 61, 62, 71–72
Eddy, John, 41, 42
Edith's checkerspot butterfly, 69–70
Egypt, 115
Egyptian culture, 4
eighteenth century, 88

Ellen, Roy, 88
emission fees. *See* pollution tax
emission reduction units (ERUs),
    129–30
enculturated, 82
Enlightenment, 87
environmental toxicology, 59
ethnocentric, 89
ethnoecology, 83
ethnography, 83–84
ethnohistory, 84
Europe, 69, 88, 91–92, 94, 96, 132, 150
European pied flycatchers, 68
European Union (EU), 128, 130, 131,
    132, 134
experimental research, 64, 72–74
externalities, 103–4
extinction, 67, 76

faculae, 42
Fagan, Brian, 86
feedback, 9–11
fish, 74–75
flexibility mechanisms, 129, 130, 134,
    137
flux models, 13–15, 25
forcing, 40
forestry, 153
fossil fuels, 151–52
    carbon dioxide (CO_2) and, 8–9
    common, 8
    defined, 7–9
    economics and, 149–50
    geologic reservoir of, 24–25
    global warming and, 27, 27–28, 84,
        98
    increasing use of, 32
    politics and, 133, 134–35, 136, 139
fossil records, 67, 76
framework convention, 126
Framework Convention on Climate
    Change (FCCC), 126, 127, 128,
    132, 133

free-air carbon dioxide enrichment
  (FACE), 73
frogs, 70
Fulani peoples, 93

"Gaia," 87
Galilei, Galileo, 41
general circulation models (GCMs),
  12–13, 47, 48
gentoo penguins, 70
geochemists, 4
geological research, 4
geologic history, 25
geologists, 3, 4
geology, 4, 5
Germany, 130, 149
glaciation, 20
glaciers, 81
glaciologists, 4
global climate models (GCMs),
  11–13, 55
global mean warming, 60
global temperature, 19–22
global warming
  Americans and, 140–41
  anthropological perspective of, 82,
    83, 84, 90, 91, 93, 96, 97, 98
  changes caused by, 148
  clean technology and, 112
  controversy, 17
  defined, 5–6
  economics and, 101–2, 103, 104,
    106, 107, 111, 114, 115, 116
  effect of trace species on, 35
  effects of, 26
  evidence for, 21–22
  Fagan on, 86
  fossil fuels and, 27–28, 84
  greenhouse-related, 62–63
  human-induced, 6, 82
  impact of, 5
  politics and, 125–26, 133
  as public good, 115
  reality of, 27

  sun and, 55
  time period of, 54–55
  as trend, 32–33
glossary, 157–66
golden toad, 70
Gore, Al, 138
great tit bird, 69
Greece, 130
greenhouse effect, 6–7, 17, 86, 92, 95,
  96
greenhouse gases (GHG)
  Annex I countries and, 129
  anthropological perspective and,
    89, 90, 98
  Canada and, 129
  defined, 7
  economics and, 101–2, 103, 105,
    106, 107, 108, 112, 113–14
  emission reduction units (ERUs),
    113–14
  in Europe, 128, 132
  global warming and, 55
  international scope of, 152
  Japan and, 129
  package of six, 128
  politics and, 121, 126, 127, 128,
    128–29, 131–32, 133–34,
    135, 136, 140–41, 142
  reversing, 153
  trifluoromethyl sulfur
    pentafluoride (SF_5CF_3), 147
  United States and, 116, 128, 132,
    151
Greenland, 94–95
Group of 77/China, 135–36

Hadley cells, 44 fig.8, 45
"hard law," 124
heterarchically, 88
historical ecology, 84, 93
Holocene epoch, 84
humans
  biological diversity and, 76
  ecology and, 67

ecosystems and, 60, 62
global warming and, 82
nature and, 87

ice ages, 4, 42, 55, 67
ice cores, 18–19, 20, 21 fig.7
ice man, 81
Iceland, 94
incentive-based (IB) regulation,
    110–11, 112
India, 114
Industrial Revolution, 8, 32, 55
infrared (IR) radiation, 7, 48
"instant law," 124–25
Intergovernmental Negotiating
    Committee (INC), 127
Intergovernmental Panel on Climate
    Change (IPCC), 125, 126, 127
international customary law, 123–24
International Negotiating Committee
    (INC), 126
international regimes, 122, 123–25,
    152–54
intertropical Convergence Zone
    (ITCZ), 45
Inuit, 94, 96
invertebrates, 70
IR spectrum, 48, 49
isotopic fractionation, 16–17
isotropic ratio, 33
Italy, 95

Japan, 117, 129, 133
joint implementation (JI), 111, 129
Judeo-Christian, 86

!Kung, 96
Kyoto Protocol
    cooperative ventures and, 153
    economics and, 106–7, 108, 111,
        114, 116
    flexibility mechanisms of, 129
    politics and, 127, 128, 129, 130–31,
        134, 135, 136–37, 140–41

requirements of, 102
United States and, 138, 140

latent heat flux, 40
latent heat of condensation/fusion, 35
Latest Paleocene Thermal Maximum
    (LPTM), 27
law, 123–25
limestone. See calcium carbonate
Linnaeus, Carolus, 86
lithosphere, 31
Little Ice Age, 42, 67
longwave (LW) radiation, 36, 38, 39,
    40, 45, 46, 47
LOWTRAN (Low Resolution
    Transmission), 49 fig.9

macrozooplankton, 71
Maglemosian culture, 96
Mali (West Africa), 89–90
Malthus, Thomas, 97
Mannion, A. M., 86
market systems, 103–5
marketable permit system, 110–11
mathematical equations, in research,
    65
Matossian, Mary, 91
Mauna Loa, 18
Maunder minima, 42
McGovern, Tom, 94
media, 145–46
Medieval Climatic Optimum, 94
Mesolithic, 96
Mesopotamia, 86
methane (CH_4), 7
methane gas hydrate, 26–27
Mexico, 153
migration, 66–67
Milton, Kay, 84
Montreal Protocol, 114, 133
morphology, 68
"Mother Nature," 87
Mount Pinatubo, 9
Muslims, 92–93

myth of primitive ecological wisdom, 89–91

NASA (National Aeronautics and Space Administration), 47
NASA Goddard Space Flight Center, 32 fig.1
Native Americans, 92
natural gas, 7
nature, 86–88
Nature Conservancy, The, 88
near infrared (NIR) radiation, 37
Netherlands, 69, 95–96, 153
New Guinea, 89
New York, 69
New Zealand, 89, 134
niche, 61, 63
nineteenth century, 87, 88
nitric acid trihydrate (NAT), 53
nitrogen, 7
nongovernmental organizations (NGOs), 140
North America, 66, 68, 69–70, 74–75
Northeast Trade Wind, 45
Northern Hemisphere, 22 fig.8, 71–72
Norway, 134, 153

observational research, 63–64
ocean ecosystems, 71
oceanographers, 4
oceans, 4, 11
oikos, 96–97
oxygen (O_2), 7, 50
ozone (O_3)
    chlorofluorocarbons (CFCs) and, 35, 114
    depletion, 31
    destruction of, 35, 114
    hole, 31–32, 53, 55
    layer, 31, 50–54, 122, 125
    role of, 48–50
    stratospheric, 31, 50–54

Pacific Ocean seabirds, 71
Paleocene epoch, 24, 27
paleoclimatologists, 4
paleoclimatology, 4
paleoecology, 66–67
paleontologists, 4
penguins, 70
Permian period, 4
phenology, 68–69
Philippines, 9
photosynthesis, 8, 24, 63, 68, 72
physiology, 68
phytoplankton, 46
plants, 72, 73
polar stratospheric cloud (PSC), 53, 54
politics, 121–42
    global climate change regime and, 125–31
    international regimes and, 122, 123–25
    time line of climate change negotiations, 127 fig.1
    United States domestic, 137–41
pollution tax, 110, 111
Polynesians, 89
Ponting, Clive, 92
population, 55, 61–62, 69–71, 97
Portugal, 130
precautionary principle, 107–8
precipitation, 71, 72, 73
preservation, 87–88, 90
protocols, 126
public goods, 115
Puerto Rico, 70

range, 69–70
ratification, of treaties, 124
religion, 91–93
Renaissance, 4
renewable energy technologies, 149
research, 17–27
    atmospheric CO_2, 18–19

carbon flux models, 22–25
correlations between climate and
    ecology, 67–72
current ecological, 66–75
ecological models in, 65–66
in ecology, approaches to, 62–66
experimental, 64, 72–74
geological, 4
global temperature, 19–22
mathematical equations in, 65
modeling techniques of, 74–75
observational, 63–64
paleoecology, 66–67
predictions, 74–75
reservoirs, 13–14
residence time, 14–15
resolutions, 124, 125
Roman Climatic Optimum, 93–94
Romanians, 92
Russia, 136
Russian Federation, 129

Scandinavia, 94
science, 150–51
Scotland, 96
Second World Climate Conference
    (Geneva, 1990), 126, 127
sedimentologists, 4
sensible heat flux, 40
shortwave (SW) radiation, 36, 37, 38,
    40, 45, 46, 47
Siberia, 81
Sierra Club, 88
sinks, 14
Siple Station (Antarctica), 34
skin cancers, 55
small pox, 92
"soft law," 124
Solar Backscatter Ultraviolet (SBUV)
    Spectrometer, 32 fig.1
solar constant, 37
solar maximum, 41
solar minimum, 41

solar radiation, 36–40, 45
sources, 14
Soviet bloc, 132, 135
species abundance, 59
species distribution, 59
stable isotopes, 16
steady state, 14
Stefan-Boltzmann Law, 38, 39
stone heads ("ahu"), 92
stratosphere, 50, 51
stratospheric ozone, 31, 50–54
Sudan, 93
sulfur dioxide (SO_2), 111
sun, 40–43, 55
sunspots, 41, 42
Sununu, John, 137
"survival of the fittest," 86–87
sustainability, 103, 103–5
Switzerland, 115, 149

Tatshenshini-Alesk Wilderness Park,
    81
technological advances, 148–49
Tertiary period, 27
Tisenjoch, 81
Total Ozone Mapping Spectrometer
    (TOMS), 32, 54 fig.11
tourism, 95–96
tradable permit system. See
    marketable permit system
transfer of technology, 154–55
treaties, 124, 131, 138, 139
trifluoromethyl sulfur pentafluoride
    (SF_5CF_3), 147
tropopause, 50
troposphere, 50, 51
tsetse fly, 93
Turkey, 92–93
twentieth century, 87

Ukraine, 129, 136
ultraviolet (UV) radiation, 37
Umbrella Group, 134

uniform emission standards, 109
United Kingdom, 68, 130
United Nations (UN), 124, 125, 126, 127
United Nations Environment Programme (UNEP), 125
United Nations General Assembly, 126
United States
    acid rain control initiative, 111
    as biggest greenhouse gas polluter, 116
    chlorofluorocarbons (CFCs) and, 52
    cloud cover, 47
    cogenerated power in, 149
    domestic politics, 137–41
    environmental regulations and, 109
    European Union (EU) and, 134–35
    fish in, 74–75
    fossil fuels and, 136, 151
    greenhouse gases (GHG), 128, 132, 133, 134–35, 151

    Kyoto Protocol and, 127, 135
    precipitation in, 71
    preservation in, 87–88
utilitarian framework, 102–3
UV-B radiation, 53–54

vertebrates, 75
Viking exploration, 4
visible region, 37
Vostok ice cores, 18–19, 20, 21 fig.7

water vapor, 35, 40, 43–46
weather prediction, 11–12
west Nile virus, 148
wildlife and fisheries management, 59
woolly mammoth, 81
World Climate Conference, 126, 127
World Climate Programme, 125
World Meteorological Organization, 125
World Trade Organization (WTO), 117